Problem Solvers

Edited by L. Marder
Senior Lecturer in Mathematics, University of Southampton

No 2

Calculus of Several Variables

Problem Solvers

Calculus of
Several Variables

L. MARDER

Senior Lecturer in Mathematics
University of Southampton

LONDON · GEORGE ALLEN & UNWIN LTD
RUSKIN HOUSE   MUSEUM STREET

First published in 1971

© George Allen & Unwin Ltd, 1971

ISBN 0 04 517001 0

Printed in Great Britain
in 10 on 12 pt 'Monophoto' Times Mathematics Series 569
by Page Bros. (Norwich) Ltd., Norwich

Contents

Chapter 1

Partial Differentiation

1.1 Definitions A *set* is any collection of objects specified by some property; the objects are called the *members* or *elements* of the set. The set of real numbers is denoted by **R**, and may be regarded as the set of points on a line (the *real axis*). A *closed interval* is a set of real numbers x satisfying a relation $a \leqslant x \leqslant b$; if the relation is changed to $a < x < b$ we obtain an *open interval*. If c is any real number, the set of points on the real axis whose Euclidean distance from c is less than δ, where $\delta > 0$, is called a *neighbourhood* of c, i.e. the relation $|x-c| < \delta$ defines a neighbourhood of c.

The set of pairs of real numbers (x, y) is denoted by $\mathbf{R}^2$, and may be regarded as the set of points in a plane. A (circular) *neighbourhood* of (a, b) is a set of points, in the plane, whose Euclidean distance from (a, b) is less than δ, where $\delta > 0$, i.e. a neighbourhood of (a, b) is defined by an inequality $(x-a)^2 + (y-b)^2 < \delta^2$. A set of points is *open* if each point P in the set possesses a neighbourhood which is entirely contained in the set. For example, the set $S: x^2 + y^2 < 1$ is open, but the set $T: x^2 + y^2 \leqslant 1$ is not, because the neighbourhoods of points $x^2 + y^2 = 1$ contain points not in T. A *boundary point* of a set is characterized by the condition that all its neighbourhoods contain both points in the set and points not in the set. The points for which $x^2 + y^2 = 1$ are boundary points for both S and T. A set such as T, which contains all its boundary points is *closed*.

A *region* is either an open set, or an open set together with some or all of its boundary points. (Usually this definition is strengthened so that a region may not consist of disjoint parts.)

A *function* is a rule of correspondence between two sets, associating one or more members of the second set with each member of the first. If the first set is $\mathbf{R}^2$ and the second is $\mathbf{R}$, then to each pair of real numbers (x, y) is associated one or more real numbers $F(x, y)$. When $z = F(x, y)$ has just one value to each pair (x, y), we call the rule (and also, rather ambiguously, the value $F(x, y)$) a *single-valued* function of the variables x and y. For example, $z = x^2 + y^2$ represents a single-valued function, whereas $z^2 = |x+y|$ is a *multi-valued* function since there correspond more than one value of z for values of x and y whose sum is not zero. Normally, we shall use the word function to imply one which is single-valued.

The variables x, y are here called the *independent* variables, z being the

1

dependent variable. The points (x, y) for which $F(x, y)$ is defined is called the *domain of definition* of the function; thus, if $F(x, y) = x^2 + y^2$ the domain of definition is the whole of the xy plane, while if $F(x, y) = \sqrt{(x-y)}$ the domain of definition is the region $x \geqslant y$. We may identify a point in three-dimensional space with each possible combination of values (x, y, z), by means of a system of rectangular cartesian coordinates $Oxyz$. In general, this *graphical representation* of a function of two variables gives rise to a surface.

Let $F(x, y)$ be defined in a neighbourhood of (a, b), except possibly at the point (a, b) itself. If we can make $F(x, y)$ arbitrarily close to a definite value l simply by choosing (x, y) to be a point sufficiently near (a, b) (but not *at* (a, b)), then we say that $F(x, y)$ *tends to the limit l as* (x, y) *tends to* (a, b). It is important that l does not depend on the direction of (x, y) from (a, b). More formally, we write

$$\lim_{(x, y) \to (a, b)} F(x, y) = l,$$

if for any given number $\epsilon > 0$, there is a number $\delta > 0$ such that

$$|F(x, y) - l| < \epsilon \quad \text{whenever} \quad 0 < (x-a)^2 + (y-b)^2 < \delta^2.$$

The function is said to be *continuous* at (a, b) if it is defined at this point, and if $F(a, b) = l$. For example, the function whose value is zero everywhere except at the point $(0, 0)$, where it has the value 1, does possess a limit as (x, y) tends to $(0, 0)$. But the limit is zero, not 1, and so the function is discontinuous at this point because $F(0, 0) \neq 0$.

Many important theorems apply to functions which are continuous at all points of a region. The sum, product, quotient, etc., of two continuous functions are all continuous, it being assumed in the latter that the denominator does not vanish. Composite functions formed only from continuous functions are themselves continuous, and so on. These theorems are generalizations of results in the calculus of one variable; precise statements will be found in most textbooks on advanced calculus.

Problem 1.1   If

$$f(x, y) = \frac{x^2(x+y)}{x^2 + y^2}, \qquad g(x, y) = \frac{x^2 - y^2 + 2x^3}{x^2 + y^2},$$

when $(x, y) \neq (0, 0)$, show that at the point $(0, 0)$: (i) f is continuous if $f(0, 0) = 0$, (ii) g is not continuous however $g(0, 0)$ may be defined.
Solution. (i) Suppose x and y are not both zero. Since $x^2 \leqslant x^2 + y^2$,

$$|f(x, y)| = \frac{x^2}{x^2 + y^2} |x + y| \leqslant |x| + |y|.$$

Therefore, if $\epsilon > 0$, we have $|f(x, y) - 0| < \epsilon$ whenever both $|x| < \frac{1}{2}\epsilon$, $|y| < \frac{1}{2}\epsilon$. This is certainly the case if $0 < x^2 + y^2 < \delta^2$, where $\delta = \frac{1}{2}\epsilon$. Therefore, $f(x, y)$ tends to zero as (x, y) tends to $(0, 0)$, and so f is continuous at this point provided $f(0, 0) = 0$.

(ii) Suppose $g(0, 0) = l$. If g is continuous at $(0, 0)$, then $g(x, y)$ must approach the value l as (x, y) approaches $(0, 0)$ along any line. But on $y = 0$, $g = 1 + 2x$ ($x \neq 0$), which approaches 1 as x approaches zero, while on $x = 0$, $g = -1$ ($y \neq 0$). The former result requires $l = 1$ and the latter $l = -1$. Since these requirements are incompatible, it follows that g cannot be continuous at the point in question. ◻

1.2 Partial Derivatives

Let $z = f(x, y)$ be a (real) function of independent (real) variables x and y. If we keep y constant, at the value y_1, then z may be regarded as a function of x. If the derivative of $z = f(x, y_1)$ with respect to x exists, for $x = x_1$, we call this the *partial derivative* of f with respect to x at the point (x_1, y_1). It is denoted variously by

$$\left.\frac{\partial f}{\partial x}\right|_{(x_1, y_1)}, \quad \text{or} \quad \left.\frac{\partial z}{\partial x}\right|_{(x_1, y_1)}, \quad \text{or} \quad f_x(x_1, y_1), \quad \text{or} \quad z_x(x_1, y_1).$$

The partial derivative with respect to y is similarly defined. Explicitly, at (x_1, y_1),

$$\frac{\partial f}{\partial x} = \lim_{h \to 0} \frac{f(x_1 + h, y_1) - f(x_1, y_1)}{h}, \tag{1.1}$$

$$\frac{\partial f}{\partial y} = \lim_{k \to 0} \frac{f(x_1, y_1 + k) - f(x_1, y_1)}{k} \tag{1.2}$$

when these limits exist. ◻

Problem 1.2 If $f(x, y) = x^2 y^3 - 2y^2$, find the values of (i) $f_x(x, y)$, (ii) $f_y(x, y)$, (iii) $f_x(-2, 1)$, (iv) $f_y(-2, 1)$.

Solution. (i) Treating y as a constant, we have on differentiation with respect to x

$$f_x(x, y) = 2xy^3.$$

(ii) Treating x as a constant and differentiating with respect to y

$$f_y(x, y) = 3x^2 y^2 - 4y.$$

On substituting $x = -2$, $y = 1$, we get

(iii) $f_x(-2, 1) = 2(-2)(1)^3 = -4$.

(iv) $f_y(-2, 1) = 3(-2)^2(1)^2 - 4(1) = 8$. ◻

Problem 1.3 Show that $z = \cos(x+y)$ is a solution of the *partial differential equation*

$$\frac{\partial z}{\partial x} - \frac{\partial z}{\partial y} = 0. \qquad (1.3)$$

Find a partial differential equation satisfied by $z = \cos xy$.

Solution. For $z = \cos(x+y)$ we have

$$\frac{\partial z}{\partial x} = \frac{\partial}{\partial x}\cos(x+y) = -\sin(x+y),$$

$$\frac{\partial z}{\partial y} = \frac{\partial}{\partial y}\cos(x+y) = -\sin(x+y),$$

whence
$$\frac{\partial z}{\partial x} - \frac{\partial z}{\partial y} = 0. \qquad (1.4)$$

For $z = \cos xy$,
$$\partial z/\partial x = -y\sin xy, \quad \partial z/\partial y = -x\sin xy,$$

so that
$$x\frac{\partial z}{\partial x} - y\frac{\partial z}{\partial y} = 0,$$

which is a partial differential equation for z (i.e. an equation involving partial derivatives of z).

Note that (1.3) is satisfied when z is an *arbitrary* function of $x+y$ and (1.4) is satisfied when z is an *arbitrary* function of xy. ☐

Problem 1.4 Give a geometrical interpretation of the partial derivatives $\partial z/\partial x$, $\partial z/\partial y$, where $z = f(x, y)$.

Solution. Consider the surface S whose rectangular cartesian equation is $z = f(x, y)$, where we take the z axis to be vertically upwards (Fig. 1.1).

Fig. 1.1

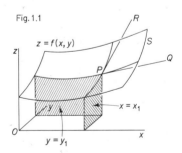

The height of the surface above any base point (x_1, y_1) in the plane $z = 0$ is $f(x_1, y_1)$, which may be positive or negative. Let P be the point $(x_1, y_1, f(x_1, y_1))$, which lies on the vertical plane curve where the plane

4

$y = y_1$ meets S. The slope of the tangent PQ to this curve, at P, in the direction of x increasing is $(\partial z/\partial x)_{(x_1, y_1)}$.

Similarly, the partial derivative $\partial z/\partial y$ at (x_1, y_1) is the slope of the tangent PR at P (in the direction y increasing) to the vertical plane curve where the plane $x = x_1$ meets S. $\qquad\square$

Problem 1.5 If
$$f(x, y) = \begin{cases} xy/(x^2 + y^2), & (x, y) \neq (0, 0), \\ 0, & (x, y) = (0, 0), \end{cases}$$
prove that f is not continuous at $(0, 0)$, but that both f_x and f_y exist at this point.

Solution. Along the line $x = cy$, $(c \neq 0)$,
$$f(x, y) = \frac{cy^2}{c^2 y^2 + y^3} = \frac{c}{c^2 + y},$$
provided $y \neq 0$. Therefore, as the point (x, y) approaches $(0, 0)$ along this line, $f(x, y)$ approaches the value $1/c$. Since this value depends on c, $f(x, y)$ does not tend to a unique limit as (x, y) tends to $(0, 0)$ (in *any* manner) and is therefore not continuous at this point.

In cases such as this, it is not reliable to differentiate the formula for f at a general point and then substitute $x = 0$, $y = 0$. Instead we work directly from the definitions (1.1), (1.2),
$$f_x(0, 0) = \lim_{h \to 0} \frac{f(h, 0) - f(0, 0)}{h} = \lim_{h \to 0} \frac{0 - 0}{h} = 0,$$
$$f_y(0, 0) = \lim_{k \to 0} \frac{f(0, k) - f(0, 0)}{k} = \lim_{k = 0} \frac{0 - 0}{k} = 0,$$
which shows that both f_x and f_y exist at $(0, 0)$, each having the value zero. $\qquad\square$

Problem 1.6 If $x = r \cos \theta$, $y = r \sin \theta$, find $\partial x/\partial r$ and $\partial r/\partial x$. Why do we not have identically
$$\frac{\partial x}{\partial r} \frac{\partial r}{\partial x} = 1?$$

Solution. If we regard r and θ as the independent variables, and differentiate the equation $x = r \cos \theta$ with respect to r (with θ constant) we obtain
$$\partial x/\partial r = \cos \theta. \qquad (1.5)$$

The notation $\partial r/\partial x$ is imprecise as it stands, since it implies that x is to be regarded as one of the independent variables but does not specify the other, i.e. it is not clear as to what we are to keep constant during the

5

differentiation. If we assume that x and y are to retain equal status in the second part of the question, then these will be the independent variables and r and θ the dependent ones. Solving the given relations we obtain

$$r = (x^2 + y^2)^{1/2}, \qquad \theta = \tan^{-1}(y/x). \qquad (1.6)$$

Differentiating the first of (1.6) with respect to x, with y constant (as the notation below indicates),

$$\left(\frac{\partial r}{\partial x}\right)_y = x(x^2 + y^2)^{-1/2} = \frac{x}{r} = \cos\theta. \qquad (1.7)$$

We may write the product of equations (1.5) and (1.7) as

$$\left(\frac{\partial x}{\partial r}\right)_\theta \left(\frac{\partial r}{\partial x}\right)_y = \cos^2\theta.$$

which is not identically equal to 1. This is to be expected since different variables are held constant in carrying out the differentiations on the left. ∎

When $f(x, y)$ possesses partial derivatives f_x and f_y in some region, these will be functions of x and y, and may themselves possess partial derivatives with respect to x and y. The latter are known as the *second partial derivatives* of f, and are written

$$f_{xx} = \frac{\partial^2 f}{\partial x^2} = \frac{\partial}{\partial x}\left(\frac{\partial f}{\partial x}\right), \qquad f_{xy} = \frac{\partial^2 f}{\partial x \partial y} = \frac{\partial}{\partial x}\left(\frac{\partial f}{\partial y}\right),$$

$$f_{yx} = \frac{\partial^2 f}{\partial y \partial x} = \frac{\partial}{\partial y}\left(\frac{\partial f}{\partial x}\right), \qquad f_{yy} = \frac{\partial^2 f}{\partial y^2} = \frac{\partial}{\partial y}\left(\frac{\partial f}{\partial y}\right).$$

Problem 1.7 Find the second partial derivatives of the function $f(x, y) = x^3 y + e^{xy}$.

Solution. We have

$$f_x = 3x^2 y + ye^{xy}, \qquad f_y = x^3 + xe^{xy}.$$

Therefore, by differentiating in accordance with the above formulae,

$$f_{xx} = 6xy + y^2 e^{xy}, \qquad f_{xy} = 3x^2 + e^{xy} + xye^{xy},$$
$$f_{yx} = 3x^2 + e^{xy} + xye^{xy}, \qquad f_{yy} = x^2 e^{xy}.$$

Note that $f_{xy} = f_{yx}$. This is not true for every function $f(x, y)$, but the relation holds, in particular, when both sides exist and are continuous near the point of interest, as is the case in most practical applications. ∎

Problem 1.8 If $f(x, y) = x^2 y^2/(x^2 + y^2)$, $(x, y) \neq (0, 0)$, prove that
(i) $xf_x + yf_y = 2f$,
(ii) $x^2 f_{xx} + 2xy f_{xy} + y^2 f_{yy} = 2f$.

6

Solution. (i) We easily find

$$f_x = 2xy^2(x^2+y^2)^{-1}+x^2y^2\frac{\partial}{\partial x}(x^2+y^2)^{-1} = 2xy^4(x^2+y^2)^{-2},$$

and by symmetry

$$f_y = 2x^4y(z^2+y^2)^{-2},$$

whence

$$xf_x+yf_y = \frac{2x^2y^2(y^2+x^2)}{(x^2+y^2)^2} = 2f. \tag{1.8}$$

(ii) Differentiating (1.8) partially with respect to x and y in turn gives

$$xf_{xx}+f_x+yf_{xy} = 2f_x,$$
$$xf_{yx}+f_y+yf_{yy} = 2f_y.$$

Multiplying the first by x, the second by y, adding, and using $f_{xy} = f_{yx}$, we get

$$x^2f_{xx}+2xyf_{xy}+y^2f_{yy} = xf_x +yf_y = 2f,$$

by (i). $\qquad\qquad\qquad\qquad\qquad\qquad\qquad\qquad\qquad\qquad\qquad\qquad$ □

Higher partial derivatives are defined as a natural extension from second derivatives. For example,

$$f_{xxx} = \frac{\partial}{\partial x}(f_{xx}), \qquad f_{yxy} = \frac{\partial}{\partial y}(f_{xy}), \qquad \text{etc.}$$

Under suitable conditions it is immaterial in which order the differentiation is performed, so that the indices may be written in any order.

1.3 Composite functions: the chain rule
Problem 1.9 If f and g are arbitrary functions of one variable, show that
$$z = f(x-ct)+g(x+ct),$$
where c is a constant, is a solution of the *wave equation*
$$\frac{\partial^2 z}{\partial x^2} = \frac{1}{c^2}\frac{\partial^2 z}{\partial t^2}.$$

Solution. Let $u = x-ct$, $v = x+ct$. Keeping t constant and applying a standard procedure for differentiating composite functions of one variable,

$$\frac{\partial z}{\partial x} = \frac{\partial}{\partial x}[f(u)+g(v)] = f'(u)\frac{\partial u}{\partial x}+g'(v)\frac{\partial v}{\partial x}$$
$$= f'(u)+g'(v),$$

where a prime denotes differentiation of a function with respect to its

7

argument. Therefore,

$$\frac{\partial^2 z}{\partial x^2} = f''(u)\frac{\partial u}{\partial x} + g''(v)\frac{\partial v}{\partial x} = f''(u) + g''(v). \tag{1.9}$$

Likewise, keeping x constant,

$$\frac{\partial z}{\partial t} = \frac{\partial}{\partial t}[f(u) + g(v)] = f'(u)\frac{\partial u}{\partial t} + g'(v)\frac{\partial v}{\partial t}$$
$$= -cf'(u) + cg'(v),$$

$$\frac{\partial^2 z}{\partial t^2} = (-c)^2 f''(u) + (c)^2 g''(v), \tag{1.10}$$

and so, by (1.9), (1.10)

$$\frac{\partial^2 z}{\partial x^2} - \frac{1}{c^2}\frac{\partial^2 z}{\partial t^2} = 0. \qquad \square$$

In general, if $w = f(x, y)$, where x and y are functions of independent variables r and s, then w is a function of r and s. We shall use the notation $\partial/\partial r$ and $\partial/\partial s$ to denote differentiation with respect to r with s constant, and differentiation with respect to s with r constant, respectively. As before, the notation $\partial/\partial x$ an $\partial/\partial y$ will imply that y and x respectively, are constant in the differentiation. The *chain rule* of partial differentiation states that

$$\frac{\partial w}{\partial r} = \frac{\partial w}{\partial x}\frac{\partial x}{\partial r} + \frac{\partial w}{\partial y}\frac{\partial y}{\partial r},$$

$$\frac{\partial w}{\partial s} = \frac{\partial w}{\partial x}\frac{\partial x}{\partial s} + \frac{\partial w}{\partial y}\frac{\partial y}{\partial s}.$$

If $w = f(x, y, z, \ldots)$, where $x = x(r, s, t, \ldots)$, $y = y(r, s, t, \ldots)$, $z = z(r, s, t, \ldots)$, etc., then the corresponding rule is

$$\frac{\partial w}{\partial r} = \frac{\partial w}{\partial x}\frac{\partial x}{\partial r} + \frac{\partial w}{\partial y}\frac{\partial y}{\partial r} + \frac{\partial w}{\partial z}\frac{\partial z}{\partial r} + \ldots,$$

$$\frac{\partial w}{\partial s} = \frac{\partial w}{\partial x}\frac{\partial x}{\partial s} + \frac{\partial w}{\partial y}\frac{\partial y}{\partial s} + \frac{\partial w}{\partial z}\frac{\partial z}{\partial s} + \ldots,$$

$$\frac{\partial w}{\partial t} = \frac{\partial w}{\partial x}\frac{\partial x}{\partial t} + \frac{\partial w}{\partial y}\frac{\partial y}{\partial t} + \frac{\partial w}{\partial z}\frac{\partial z}{\partial t} + \ldots, \quad \text{etc.}$$

Here we suppose that the numbers of variables $x, y, z, \ldots$ and $r, s, t, \ldots$ are both finite though not necessarily the same.

Problem 1.10 Let $w = f(x, y) = e^{x(x-y)}$, where $x = 2t\cos t$, $y = 2t\sin t$. Find dw/dt when $t = \pi$.

8

Solution. We can either substitute for x and y in terms of t, and then carry out the differentiation, or we can apply the chain rule, obtaining (since w is a composite function of t only)

$$\frac{dw}{dt} = \frac{\partial f}{\partial x}\frac{dx}{dt} + \frac{\partial f}{\partial y}\frac{dy}{dt}$$

$$= (2x-y)e^{x(x-y)}(2\cos t - 2t\sin t) - xe^{x(x-y)}(2\sin t + 2t\cos t).$$

When $t = \pi$, $x = -2\pi$, $y = 0$, so that

$$dw/dt = (-4\pi)e^{4\pi^2}(-2) - (-2\pi)e^{4\pi^2}(-2\pi) = 4\pi(2-\pi)e^{4\pi^2}. \qquad \square$$

Problem 1.11 If $x = \rho\cos\theta$, $y = \rho\sin\theta$ (ρ, θ being plane polar coordinates and x, y rectangular coordinates), show that Laplace's equation for $V(x, y)$,

$$\frac{\partial^2 V}{\partial x^2} + \frac{\partial^2 V}{\partial y^2} = 0,$$

is equivalent to

$$\frac{\partial^2 V}{\partial \rho^2} + \frac{1}{\rho}\frac{\partial V}{\partial \rho} + \frac{1}{\rho^2}\frac{\partial^2 V}{\partial \theta^2} = 0.$$

Solution. First method. We have

$$\frac{\partial V}{\partial \rho} = \frac{\partial V}{\partial x}\frac{\partial x}{\partial \rho} + \frac{\partial V}{\partial y}\frac{\partial y}{\partial \rho} = \cos\theta\frac{\partial V}{\partial x} + \sin\theta\frac{\partial V}{\partial y}, \qquad (1.11)$$

$$\frac{\partial V}{\partial \theta} = \frac{\partial V}{\partial x}\frac{\partial x}{\partial \theta} + \frac{\partial V}{\partial y}\frac{\partial y}{\partial \theta} = -\rho\sin\theta\frac{\partial V}{\partial x} + \rho\cos\theta\frac{\partial V}{\partial y}. \qquad (1.12)$$

Thus, $\partial/\partial\rho$ and $\partial/\partial\theta$ may be replaced by equivalent operations

$$\frac{\partial}{\partial \rho} = \cos\theta\frac{\partial}{\partial x} + \sin\theta\frac{\partial}{\partial y}, \qquad (1.13)$$

$$\frac{\partial}{\partial \theta} = -\rho\sin\theta\frac{\partial}{\partial x} + \rho\cos\theta\frac{\partial}{\partial y}. \qquad (1.14)$$

By (1.11),

$$\frac{\partial^2 V}{\partial \rho^2} = \frac{\partial}{\partial \rho}\left(\cos\theta\frac{\partial V}{\partial x} + \sin\theta\frac{\partial V}{\partial y}\right)$$

$$= \cos\theta\frac{\partial}{\partial \rho}\left(\frac{\partial V}{\partial x}\right) + \frac{\partial V}{\partial x}\frac{\partial}{\partial \rho}(\cos\theta)$$

$$+ \sin\theta\frac{\partial}{\partial \rho}\left(\frac{\partial V}{\partial y}\right) + \frac{\partial V}{\partial y}\frac{\partial}{\partial \rho}(\sin\theta). \qquad (1.15)$$

Now use (1.13) to replace $\partial/\partial\rho$ in the first and third terms only; the differentiation with respect to ρ in the other two terms can be carried out

9

directly. (In fact, since ρ and θ are independent variables, we have $(\partial/\partial\rho)\cos\theta = 0$, $(\partial/\partial\rho)\sin\theta = 0$.) Therefore,

$$\frac{\partial^2 V}{\partial\rho^2} = \cos\theta\left(\cos\theta\frac{\partial}{\partial x}+\sin\theta\frac{\partial}{\partial y}\right)\frac{\partial V}{\partial x}+\sin\theta\left(\cos\theta\frac{\partial}{\partial x}+\sin\theta\frac{\partial}{\partial y}\right)\frac{\partial V}{\partial y}$$

$$= \cos^2\theta\frac{\partial^2 V}{\partial x^2}+2\sin\theta\cos\theta\frac{\partial^2 V}{\partial x\partial y}+\sin^2\theta\frac{\partial^2 V}{\partial y^2}, \tag{1.16}$$

since $\partial^2 V/\partial x\partial y = \partial^2 V/\partial y\partial x$. Similarly,

$$\frac{\partial^2 V}{\partial\theta^2} = \frac{\partial}{\partial\theta}\left(-\rho\sin\theta\frac{\partial V}{\partial x}+\rho\cos\theta\frac{\partial V}{\partial y}\right).$$

But

$$\frac{\partial}{\partial\theta}(-\rho\sin\theta) = -\rho\cos\theta, \quad \frac{\partial}{\partial\theta}(\rho\cos\theta) = -\rho\sin\theta,$$

and by the procedure adopted for (1.15) we obtain, using (1.14),

$$\frac{\partial^2 V}{\partial\theta^2} = -\rho\sin\theta\left(-\rho\sin\theta\frac{\partial}{\partial x}+\rho\cos\theta\frac{\partial}{\partial y}\right)\frac{\partial V}{\partial x}-\rho\cos\theta\frac{\partial V}{\partial x}$$

$$+\rho\cos\theta\left(-\rho\sin\theta\frac{\partial}{\partial x}+\rho\cos\theta\frac{\partial}{\partial y}\right)\frac{\partial V}{\partial y}-\rho\sin\theta\frac{\partial V}{\partial y}$$

$$= \rho^2\left(\sin^2\theta\frac{\partial^2 V}{\partial x^2}-2\sin\theta\cos\theta\frac{\partial^2 V}{\partial x\partial y}+\cos^2\theta\frac{\partial^2 V}{\partial y^2}\right)$$

$$-\rho\left(\cos\theta\frac{\partial V}{\partial x}+\sin\theta\frac{\partial V}{\partial y}\right). \tag{1.17}$$

By (1.11), (1.15), (1.17),

$$\frac{\partial^2 V}{\partial\rho^2}+\frac{1}{\rho}\frac{\partial V}{\partial\rho}+\frac{1}{\rho^2}\frac{\partial^2 V}{\partial\theta^2} = \frac{\partial^2 V}{\partial x^2}+\frac{\partial^2 V}{\partial y^2} \tag{1.18}$$

whence the required result follows.

Second method. *Either* by inverting the equations $x = \rho\cos\theta$, $y = \rho\sin\theta$, to obtain $\rho = (x^2+y^2)^{1/2}$, $\theta = \tan^{-1}(y/x)$, *or* by solving (1.13), (1.14), we find that

$$\frac{\partial}{\partial x} = \cos\theta\frac{\partial}{\partial\rho}-\frac{1}{\rho}\sin\theta\frac{\partial}{\partial\theta}, \quad \frac{\partial}{\partial y} = \sin\theta\frac{\partial}{\partial\rho}+\frac{1}{\rho}\cos\theta\frac{\partial}{\partial\theta}. \tag{1.19}$$

By applying, in turn, each of these operators to V we can obtain expressions for $\partial^2 V/\partial x^2$ and $\partial^2 V/\partial y^2$ in terms of ρ,θ and derivatives of V (up to second order) with respect to these variables. On adding the two expressions so obtained, we get, after a little reduction, the left-hand side of (1.18), so that the required result again follows. The reader should fill in the details.

Note that it is better not to rely on inverting given relations in problems of this type; this may not be possible. The use of (1.13), (1.14) to derive (1.19) is better. The method is illustrated further in the next problem. □

Problem 1.12 If $x = u^2 - v^2$, $y = 2uv$, find $\partial u/\partial x$, $\partial v/\partial x$, $\partial u/\partial y$, $\partial v/\partial y$. If $f = f(x, y)$, express $(\partial f/\partial x)^2 + (\partial f/\partial y)^2$ in terms of the partial derivatives of f with respect to u and v.

Solution. According to the chain rule, for any function $g(x, y)$

$$\frac{\partial g}{\partial x} = \frac{\partial g}{\partial u}\frac{\partial u}{\partial x} + \frac{\partial g}{\partial v}\frac{\partial v}{\partial x}, \tag{1.20}$$

$$\frac{\partial g}{\partial y} = \frac{\partial g}{\partial u}\frac{\partial u}{\partial y} + \frac{\partial g}{\partial v}\frac{\partial v}{\partial y}. \tag{1.21}$$

In particular, when $g \equiv x$, we find from the given relations

$$1 = 2u\frac{\partial u}{\partial x} - 2v\frac{\partial v}{\partial x},$$

$$0 = 2u\frac{\partial u}{\partial y} - 2v\frac{\partial v}{\partial y}.$$

Similarly, when $g \equiv y$,

$$0 = 2v\frac{\partial u}{\partial x} + 2u\frac{\partial v}{\partial x},$$

$$1 = 2v\frac{\partial u}{\partial y} + 2u\frac{\partial v}{\partial y}.$$

Solving the last four equations,

$$\frac{\partial u}{\partial x} = \frac{u}{2(u^2 + v^2)}, \qquad \frac{\partial v}{\partial x} = \frac{-v}{2(u^2 + v^2)},$$

$$\frac{\partial u}{\partial y} = \frac{v}{2(u^2 + v^2)}, \qquad \frac{\partial v}{\partial y} = \frac{u}{2(u^2 + v^2)}.$$

Substituting these values in (1.20), (1.21) and replacing g by f:

$$\frac{\partial f}{\partial x} = \frac{1}{2(u^2 + v^2)}\left(u\frac{\partial f}{\partial u} - v\frac{\partial f}{\partial v}\right), \tag{1.22}$$

$$\frac{\partial f}{\partial y} = \frac{1}{2(u^2 + v^2)}\left(v\frac{\partial f}{\partial u} + u\frac{\partial f}{\partial v}\right), \tag{1.23}$$

whence
$$\left(\frac{\partial f}{\partial x}\right)^2 + \left(\frac{\partial f}{\partial y}\right)^2 = \frac{1}{4(u^2 + v^2)}\left[\left(\frac{\partial f}{\partial u}\right)^2 + \left(\frac{\partial f}{\partial v}\right)^2\right].$$ □

Problem 1.13   In the notation of the last problem, find $\partial^2 f/\partial x\partial y$ in terms of the derivatives of f with respect to u and v, when $u = 2, v = 1$.

Solution. From (1.22) we deduce that

$$\frac{\partial}{\partial x} = \frac{1}{2(u^2+v^2)}\left(u\frac{\partial}{\partial u} - v\frac{\partial}{\partial v}\right),$$

and applying this operator to (1.23) we have, when $u = 2, v = 1$,

$$\frac{\partial^2 f}{\partial x\partial y} = \frac{1}{10}\left(2\frac{\partial}{\partial u} - \frac{\partial}{\partial v}\right)\left[\frac{1}{2(u^2+v^2)}\left(v\frac{\partial f}{\partial u} + u\frac{\partial f}{\partial v}\right)\right]$$

$$= \frac{1}{20}\left[\left(2\frac{\partial}{\partial u} - \frac{\partial}{\partial v}\right)\frac{1}{u^2+v^2}\right]\left(\frac{\partial f}{\partial u} + 2\frac{\partial f}{\partial v}\right)$$

$$+ \frac{1}{100}\left(2\frac{\partial}{\partial u} - \frac{\partial}{\partial v}\right)\left(v\frac{\partial f}{\partial u} + u\frac{\partial f}{\partial v}\right).$$

Note that we have replaced u by 1, and v by 2, only in those functions of u and v which are not still to be differentiated.

Carrying out the differentiations, and remembering that u and v are independent variables, we find

$$\frac{\partial^2 f}{\partial x\,\partial y} = \frac{1}{20}\left[\frac{-4u+2v}{(u^2+v^2)^2}\right]\left(\frac{\partial f}{\partial u} + 2\frac{\partial f}{\partial v}\right)$$

$$+ \frac{1}{100}\left[2v\frac{\partial^2 f}{\partial u^2} - v\frac{\partial^2 f}{\partial v\,\partial u} - \frac{\partial f}{\partial u} + 2u\frac{\partial^2 f}{\partial u\,\partial v} + 2\frac{\partial f}{\partial v} - u\frac{\partial^2 f}{\partial v^2}\right]$$

$$= \frac{1}{500}\left(10\frac{\partial^2 f}{\partial u^2} + 15\frac{\partial^2 f}{\partial u\,\partial v} - 10\frac{\partial^2 f}{\partial v^2} - 11\frac{\partial f}{\partial u} - 2\frac{\partial f}{\partial v}\right). \qquad \square$$

Problem 1.14   Find the general solution of the wave equation for $z(x, t)$:

$$\frac{\partial^2 z}{\partial x^2} - \frac{1}{c^2}\frac{\partial^2 z}{\partial t^2} = 0. \tag{1.24}$$

Solution. Let ∂_x, ∂_y denote $\partial/\partial x, \partial/\partial y$, and put $u = x+ct, v = x-ct$. (This substitution is suggested by Problem 1.9). Then, by the method of the foregoing problems, we have

$$\partial_x = u_x\partial_u + v_x\partial_v = \partial_u + \partial_v,$$

$$\partial_t = u_t\partial_u + v_t\partial_v = c(\partial_u - \partial_v),$$

and so, in an obvious notation for second derivatives,

$$\partial_{xx} = (\partial_u + \partial_v)(\partial_u + \partial_v) = \partial_{uu} + 2\partial_{uv} + \partial_{vv},$$

$$\frac{1}{c^2}\partial_{tt} = (\partial_u - \partial_v)(\partial_u - \partial_v) = \partial_{uu} - 2\partial_{uv} + \partial_{vv}.$$

By subtracting, we find that (1.24) may be written
$$\partial^2 z/\partial u\, \partial v = 0.$$
Integration with respect to u with v constant gives
$$\partial z/\partial v = F(v),$$
where F is an arbitrary function. Next, integration with respect to v gives
$$z = \int F(v)\, dv + f(u),$$
where f is arbitrary. Therefore on writing $g(v)$ for the indefinite integral in the last equation we obtain the general solution of (1.24), involving two arbitrary functions f and g:
$$z = f(x+ct) + g(x-ct). \qquad \square$$

Problem 1.15 Find the solution of the *spherical wave equation* for $z(r,t)$:
$$\frac{\partial^2 z}{\partial r^2} + \frac{2}{r}\frac{\partial z}{\partial r} - \frac{1}{c^2}\frac{\partial^2 z}{\partial t^2} = 0, \tag{1.25}$$
such that $z = r^{-1}\sin r$, and $\partial z/\partial t = 0$, when $t = 0$.
  Solution. Since
$$\partial_{rr}(rz) = \partial_r(rz_r + z) = rz_{rr} + 2z_r,$$
$$\partial_{tt}(rz) = \partial_t(rz_t) = rz_{tt},$$
we find that (1.25) can be written, on multiplying by r,
$$[\partial_{rr} - (1/c^2)\partial_{tt}](rz) = 0.$$
By the last problem, the general solution of this equation is
$$rz = f(r+ct) + g(r-ct), \tag{1.26}$$
where f and g are arbitrary. Using the given conditions at $t = 0$, we obtain
$$(z)_{(t=0)} = r^{-1}\sin r = r^{-1}[f(r) + g(r)]$$
i.e.
$$f(r) + g(r) = \sin r, \tag{1.27}$$
and
$$(\partial_t z)_{(t=0)} = 0 = r^{-1}[f'(r+ct)(c) + g'(r-ct)(-c)]_{(t=0)}$$
or
$$f'(r) - g'(r) = 0, \tag{1.28}$$
by differentiation of (1.26), where a prime denotes differentiation of a function with respect to its argument.
  Note that (1.28) does not involve t, and the derivatives occurring there are ordinary and not partial derivatives. Hence we may integrate to obtain
$$f(r) - g(r) = A = \text{const.} \tag{1.29}$$
By (1.27), (1.29),
$$f(r) = \tfrac{1}{2}(\sin r + A), \qquad g(r) = \tfrac{1}{2}(\sin r - A),$$

13

and therefore, the required solution is by (1.26),

$$z = r^{-1}[f(r+ct)+g(r-ct)]$$
$$= \tfrac{1}{2}[\sin(r+ct)+\sin(r-ct)]. \qquad \square$$

If the function $f(x, y, z, \ldots)$ has the property

$$f(tx, ty, tz, \ldots) = t^n f(x, y, z, \ldots),$$

for general values of t, it is said to be *homogeneous of degree n*. For example,

$$f(x, y, z) = (x^3 + 3xy^2 - xz^2)/z$$

is homogeneous of degree 2 because, if $t \neq 0$,

$$f(tx, ty, tz) = [(tx)^3 + 3(tx)(ty)^2 - (tx)(tz)^2]/tz$$
$$= t^2(x^3 + 3xy^2 - xz^2)/z = t^2 f(x, y, z).$$

Problem 1.16 Prove *Euler's theorem* that if $f(x, y, z, \ldots)$ is homogeneous of degree n, then

$$x\frac{\partial f}{\partial x} + y\frac{\partial f}{\partial y} + z\frac{\partial f}{\partial z} + \ldots = nf. \qquad (1.30)$$

Verify the result in the case $f(x, y, z) = 2x^3 + yz^2 - xyz$.

Solution. Let $X = tx$, $Y = ty$, $Z = tz, \ldots$. Then we are given

$$f(X, Y, Z, \ldots) = t^n f(x, y, z, \ldots).$$

Differentiate each side with respect to t, to get

$$f_X\frac{dX}{dt} + f_Y\frac{dY}{dt} + f_Z\frac{dZ}{dt} + \ldots = nt^{n-1}f(x, y, z, \ldots),$$

i.e.
$$xf_X + yf_Y + zf_Z + \ldots = nt^{n-1}f(x, y, z, \ldots), \qquad (1.31)$$

where $f_X = (\partial/\partial X)f(X, Y, Z, \ldots)$, etc.

Now put $t = 1$, so that $X = x$, $Y = y$, $Z = z, \ldots$, and obtain from (1.31)

$$xf_x + yf_y + zf_z + \ldots = nf,$$

as required.

The given function is evidently homogeneous of degree 3. By direct calculation,

$$xf_x + yf_y + zf_z = x(6x^2 - yz) + y(z^2 - xz) + z(2yz - xy)$$
$$= 6x^3 - 3xyz + 3yz^2 = 3f,$$

which was to be proved. $\qquad \square$

A generalization of Euler's theorem states that (c.f. Problem 1.8)

$$(x\partial_x + y\partial_y + \ldots)^m f(x, y, \ldots) = n(n-1)\ldots(n-m+1)f(x, y, \ldots),$$

for $m \leqslant n$.

14

1.4 Differentials   Let P be the point (x_1, y_1) and Q the point $(x_1 + \Delta x, y_1 + \Delta y)$, where Δx and Δy are increments in x and y respectively, and let R be a variable point $(x_1 + t\,\Delta x, y_1 + t\,\Delta y)$ on the straight line PQ, t being a parameter. If P and Q are fixed, then for any function $f(x, y)$

$$F(t) \equiv f(x_1 + t\,\Delta x, y_1 + t\,\Delta y) \tag{1.32}$$

is a function of t only. By a mean-value theorem of the calculus of one variable, if F has a derivative in $0 \leqslant t \leqslant 1$, then

$$F(1) - F(0) = F'(k),$$

for some k such that $0 < k < 1$. This quantity represents the increment Δf in f between P and Q, and by differentiation of (1.32) we obtain

$$\Delta f = f_x(x_1 + k\,\Delta x, y_1 + k\,\Delta y)\,\Delta x + f_y(x_1 + k\,\Delta x, y_1 + k\,\Delta y)\,\Delta y,$$

assuming these derivatives exist. If, further, they are continuous, we can write the last equation as

$$\Delta f = (f_x + \epsilon_1)\,\Delta x + (f_y + \epsilon_2)\,\Delta y,$$

where the partial derivatives are now evaluated at (x_1, y_1) and ϵ_1 and ϵ_2 both tend to zero as Δx and Δy tend to zero.

The *principal part* of this increment is

$$df = f_x\,\Delta x + f_y\,\Delta y,$$

and when Δx and Δy are small, df is approximately equal to Δf. We call df the *differential* of f, and when the increments Δx and Δy are used in this connection they are denoted by dx and dy, respectively. Thus,

$$df = f_x\,dx + f_y\,dy. \tag{1.33}$$

This formula holds even when x and y are not the independent variables, in which case the meanings of dx and dy change slightly (because they are now principal parts). For example, if $x = x(u, v)$, $y = y(u, v)$, where u and v are independent variables, then corresponding to differentials du, dv in u, v, we have

$$dx = x_u\,du + x_v\,dv, \quad dy = y_u\,du + y_v\,dv, \quad df = f_u\,du + f_v\,dv.$$

These satisfy (1.33) because

$$f_x x_u + f_y y_u = f_u, \qquad f_x x_v + f_y y_v = f_v.$$

There are immediate generalizations to functions of more than two variables.

Problem 1.17   (i) Find df when $f(x, y) = x^2 e^{2y} \cos xy$; (ii) find $g(x, y)$ such that

$$dg = [2y^2(\sin x + x \cos x) - ye^{xy}]\,dx + (4xy \sin x - xe^{xy} + 2y)\,dy. \tag{1.34}$$

Solution. (i) By (1.33),

15

$$df = e^{2y}[2x\cos xy + x^2(-y)\sin xy]\,dx + x^2[2e^{2y}\cos xy + e^{2y}(-x)\sin xy]\,dy$$
$$= xe^{2y}[(2\cos xy - xy\sin xy)\,dx + x(2\cos xy - x\sin xy)\,dy].$$

(ii) If such a function $g(x, y)$ exists, then by comparing (1.34) with (1.33) (g replacing f), we find that g must satisfy the conditions

$$g_x = 2y^2(\sin x + x\cos x) - ye^{xy}, \tag{1.35}$$
$$g_y = 4xy\sin x - xe^{xy} + 2y. \tag{1.36}$$

Integrate (1.35) with respect to x, keeping y constant, to get

$$g = 2y^2x\sin x - e^{xy} + h(y),$$

where the function h is to be determined. Substitute in (1.36):

$$g_y = 4xy\sin x - xe^{xy} + h'(y) = 4xy\sin x - xe^{xy} + 2y,$$

whence we must have $h'(y) = 2y$, i.e.

$$h(y) = y^2 + C,$$

where C is an arbitrary constant. Thus,

$$g(x, y) = 2y^2x\sin x - e^{xy} + y^2 + C. \qquad \square$$

Problem 1.18 Show that there is no function (having continuous second partial derivatives) whose differential is $xy\,dx + 2x^2\,dy$.

Solution. Consider any function $f(x, y)$ having the differential

$$df = P(x, y)\,dx + Q(x, y)\,dy, \tag{1.37}$$

where P and Q are given functions. By (1.33), we must have

$$f_x = P, \qquad f_y = Q.$$

When f possesses continuous second partial derivatives, then $f_{yx} = f_{xy}$, so that

$$\partial P/\partial y = \partial Q/\partial x \tag{1.38}$$

is a necessary condition for the right-hand side of (1.37) to be the differential of some such function f. In the present case, $P = xy$, $Q = 2x^2$, and

$$P_y = x, \qquad Q_x = 2x,$$

i.e.
$$P_y \neq Q_x$$

(except when $x = 0$) and so the statement in the question is proved.

When (1.37) holds, for some function $f(x, y)$, the expression $P\,dx + Q\,dy$ is said to be an *exact* differential. A converse result to that derived here is that when (1.38) holds throughout a region the differential form $P\,dx + Q\,dy$ is exact, (although f may not be single-valued unless the region is *simply-connected*; see p. 72). $\qquad \square$

Problem 1.19 Give geometrical interpretations of Δf and df at the point (x_1, y_1) for a given function $f(x, y)$.

16

Solution. Consider the surface $S: z = f(x, y)$ referred to rectangular cartesian axes $Oxyz$. Let P_0 and Q_0 denote the points (x_1, y_1) and (x_2, y_2) (the latter being chosen arbitrarily) in the xy plane, and put $z_1 = f(x_1, y_1)$, $z_2 = f(x_2, y_2)$, so that the points $P(x_1, y_1, z_1)$, $Q(x_2, y_2, z_2)$ lie on S. If $dx = \Delta x = x_2 - x_1$, $dy = \Delta y = y_2 - y_1$, the corresponding increment in f is

$$\Delta f = z_2 - z_1 = f(x_1 + \Delta x, y_1 + \Delta y) - f(x_1, y_1),$$

i.e. Δf is the increase in height of the surface S above the base point (x, y) as the latter moves from P_0 to Q_0.

Consider next the tangent plane to S at the point $P(x_1, y_1, z_1)$, whose equation, being linear in x, y, z, must be of the form

$$z - z_1 = l(x - x_1) + m(y - y_1). \tag{1.39}$$

By inspection, l is the slope of the line of intersection of this plane and the vertical plane $y = y_1$. The curve of intersection of S and the plane $y = y_1$ must have the same slope, l, at P, and so

$$l = f_x(x_1, y_1), \qquad m = f_y(x_1, y_1), \tag{1.40}$$

(the second result following from a similar argument).

If we substitute x_2, y_2 for x, y, then by (1.40) the right-hand side of (1.39) becomes

$$f_x\, dx + f_y\, dy, \tag{1.41}$$

i.e. the differential df, evaluated at (x_1, y_1). The left-hand side of (1.39) becomes

$$z_2' - z_1, \tag{1.42}$$

where z_2' is the height vertically above Q_0 of the tangent plane at P. Equating (1.41), (1.42) we get

$$df = z_2' - z_1,$$

which shows that the differential df is the increase in height of the *tangent plane* at P as the base point (x, y) moves from P_0 to Q_0.

This result further illustrates the approximate equality between df and Δf when dx and dy are small. $\qquad\square$

Problem 1.20 If $f(x, y) = e^{x^2 y}$, find an approximate value for $f(1 \cdot 05, 0 \cdot 97)$.

Solution. Write $x = x_0 + \Delta x$, $y = y_0 + \Delta y$, where x_0 and y_0 are arbitrary. If Δx and Δy are sufficiently small we have approximately

$$f(x, y) - f(x_0, y_0) = f_x(x_0, y_0)\,\Delta x + f_y(x_0, y_0)\,\Delta y.$$

Choose $x_0 = 1$, $y_0 = 1$, $\Delta x = 0 \cdot 05$, $\Delta y = -0 \cdot 03$. Then $f(x_0, y_0) = e$, and

$$f_x(x_0, y_0) = 2x_0 e^{x_0^2 y_0} = 2e,$$
$$f_y(x_0, y_0) = x_0^2 e^{x_0^2 y_0} = e.$$

Therefore the required approximate value is

$$f(1{\cdot}05, 0{\cdot}97) = e + 2e(0{\cdot}05) + e(-0{\cdot}03)$$
$$= 1{\cdot}07e = 2{\cdot}91.$$

The exact value to 4 decimal places is, in fact, $2{\cdot}9137$.

Closer approximations can be found using Taylor's theorem; see Problem 3.4. $\square$

Problem 1.21 If a, b, c are the sides of a triangle, express a in terms of b, c and the opposite angle A; and determine the differential da. Hence find an approximate value for a when $b = 4{\cdot}10, c = 3{\cdot}95, A = 62°$.

Solution. By the cosine formula,

$$a^2 = b^2 + c^2 - 2bc \cos A. \tag{1.43}$$

Taking differentials,

$$2a\,da = 2b\,db + 2c\,dc - 2c \cos A\,db - 2b \cos A\,dc + 2bc \sin A\,dA,$$

i.e. $\quad da = a^{-1}[(b - c \cos A)\,db + (c - b \cos A)\,dc + bc \sin A\,dA]. \tag{1.44}$

For the required approximation, start with the values $b = 4$, $c = 4$, $A = 60°$, so that by (1.43)

$$a^2 = (4)^2 + (4)^2 - 2(4)(4)(\tfrac{1}{2}) = 16, \quad \text{i.e. } a = 4$$

(which also follows since the starting values correspond to an equilateral triangle). Now put

$$\Delta b = db = 0{\cdot}10, \quad \Delta c = dc = -0{\cdot}05,$$
$$\Delta A = dA = (62 - 60)\,\pi/180 = \pi/90$$

(radians), and obtain an approximate relation from (1.44) by replacing da by Δa, giving

$$\Delta a = \tfrac{1}{4}[4 - 4(\tfrac{1}{2})](0{\cdot}10) + [4 - 4(\tfrac{1}{2})](-0{\cdot}05) + 4(4)(\tfrac{1}{2}\sqrt{3})(\pi/90)$$
$$= \tfrac{1}{4}(0{\cdot}20 - 0{\cdot}10 + 0{\cdot}484) = 0{\cdot}15,$$

to two decimal places.

Thus, the approximate value for a is $4{\cdot}15$. This, in fact, agrees with the exact value taken to two decimal places. $\square$

EXERCISES

1.   If $f(x, y) = x \cos y + y \sin xy$, find $f_x(x, y), f_y(x, y), f_{xx}(x, y), f_{xx}(1, \tfrac{1}{2}\pi)$, $f_{xy}(1, \tfrac{1}{2}\pi)$.

2.   If $f(x, y) = xy(x^2 - y^2)/(x^2 + y^2)$ when $(x, y) \neq (0, 0)$, and $f(0, 0) = 0$, show that f is continuous at $(0, 0)$. Show from the definitions that $f_{xy}(0, 0)$ and $f_{yx}(0, 0)$ both exist, but have different values.

18

3. If $z = \sin xy^2$, and $x = t+e^t$, $y = te^{-t}$, find dz/dt and d^2z/dt^2 when $t = 1$.

4. If $x = e^u \cos v$, $y = e^u \sin v$, find $\partial u/\partial x$, $\partial u/\partial y$, $\partial v/\partial x$, $\partial v/\partial y$ in terms of x and y, (i) by the elimination method of Problem 1.12, (ii) by first solving for u and v. Show that for an arbitrary function $f(x, y)$ (which is continuous with continuous partial derivatives up to second order),

$$\frac{\partial^2 f}{\partial x^2} + \frac{\partial^2 f}{\partial y^2} = e^{-2u}\left[\frac{\partial^2 f}{\partial u^2} + \frac{\partial^2 f}{\partial v^2}\right].$$

5. Show that $z = f(x+e^y) + g(x-e^y)$ is a solution of the equation

$$e^{2y}\frac{\partial^2 z}{\partial x^2} - \frac{\partial^2 z}{\partial y^2} + \frac{\partial z}{\partial y} = 0$$

for arbitrary f and g (possessing continuous second derivatives). Find a solution such that $z = 0$, $\partial z/\partial y = 1+x$, when $y = 0$.

6. Show that $f(x, y) = \sqrt{(x^4 + y^4)}$ satisfies the equation

$$xf_x + yf_y = 2f,$$

(i) by calculation of the partial derivatives, (ii) by Euler's theorem on homogeneous functions.
Show also that

$$x^2 f_{xx} + 2xy f_{xy} + y^2 f_{yy} = 2f.$$

7. (i) Show that $(xy+1)y\,dx + [(xy-1)x+y]\,dy$ is not an exact differential. (ii) Find $f(x, y)$ such that

$$df = y[(1+x)e^{x+y} - 2x]\,dx + x[(1+y)e^{x+y} - x]\,dy.$$

8. Find an approximate value, to two decimal places, for

$$\ln [(0.97)e^{0.48} - (0.03)e^{0.52}]$$

by considering the differential of $\ln[(1-x)e^{1-y} - xe^y]$ at $x = 0$, $y = 0.5$.

Chapter 2

Jacobians and Transformations

2.1 Implicit Functions and Jacobians When three variables x, y, z are related by a single equation of the form

$$F(x, y, z) = 0, \tag{2.1}$$

we can usually assign values to two of them, x and y say, and the third variable, z, is then determined. We say that (2.1) defines z as an *implicit* function of x and y.

But not every equation of this form defines any one variable (supposed real) as an implicit function of the other two. For example, the equation

$$x^2 + y^2 + z^2 + 1 = 0$$

is never satisfied when all the variables take real values. Again, the equation

$$(x + y)^2 + z^2 - 1 = 0 \tag{2.2}$$

defines a real z only when $(x+y)^2 \leqslant 1$, and at each point not on the boundary in this domain of dependence the function is multi-valued.

When (2.1) does define z as a function of x and y in a region R of the xy plane, $z = f(x, y)$ say, then (2.1) becomes an identity in x and y if $f(x, y)$ is substituted for z. Thus, if F is the function on the left in (2.2), then in the strip R whose equation is $|x+y| \leqslant 1$, we have

$$z = f(x, y) = \pm\sqrt{[1-(x+y)^2]},$$

and on substituting this expression for z in (2.2) we get

$$F[(x, y, f(x, y)] = (x+y)^2 + 1 - (x+y)^2 - 1 = 0,$$

which is an identity.

Suppose that one set of values x_0, y_0, z_0 can be found which satisfy (2.1), and that near (x_0, y_0, z_0), F and its first partial derivatives are continuous and $\partial F/\partial z \neq 0$. Then an *existence theorem* states that in a certain region of the xy plane containing (x_0, y_0), there is precisely one differentiable function $z = f(x, y)$ which reduces (2.1) to an identity and is such that $z_0 = f(x_0, y_0)$.

Problem 2.1 If $x^2 - xz + z^2 + yz = 4$, find $\partial z/\partial x$ and $\partial z/\partial y$ when $x = 1$, $y = 3$.

Solution. We have

$$F(x, y, z) \equiv x^2 - xz + z^2 + yz - 4 = 0. \tag{2.3}$$

20

If we assume that z is defined as a function of x and y by this equation, and differentiate with respect to x (with y constant), we get

$$2x - z - x\frac{\partial z}{\partial x} + 2z\frac{\partial z}{\partial x} + y\frac{\partial z}{\partial x} = 0. \tag{2.4}$$

Similarly, differentiating with respect to y (x constant),

$$-x\frac{\partial z}{\partial y} + 2z\frac{\partial z}{\partial y} + y\frac{\partial z}{\partial y} + z = 0. \tag{2.5}$$

Solving (2.4), (2.5) gives

$$\frac{\partial z}{\partial x} = \frac{z - 2x}{2z + y - x}, \qquad \frac{\partial z}{\partial y} = \frac{-z}{2z + y - x}. \tag{2.6}$$

When $x = 1$, $y = 3$, (2.3) gives

$$z^2 + 2z - 3 = 0,$$

i.e. $\qquad\qquad\qquad z = 1 \text{ or } -3. \tag{2.7}$

Substituting from (2.7) in (2.6),

$$\frac{\partial z}{\partial x} = -\frac{1}{4} \text{ or } \frac{5}{4}, \qquad \frac{\partial z}{\partial y} = -\frac{1}{4} \text{ or } -\frac{3}{4};$$

the first value in each case corresponding to $z = 1$ and the second to $z = -3$. □

Problem 2.2 If u and v are functions of x and y, defined implicitly in some region of the xy plane by the equations

$$u \sin v + x^2 = 0,$$
$$u \cos v - y^2 = 0,$$

find $\partial u/\partial x$, $\partial v/\partial x$, $\partial u/\partial y$, $\partial v/\partial y$.

Solution. Differentiating the given equations with respect to x, with y constant,

$$\frac{\partial u}{\partial x}\sin v + u \cos v \frac{\partial v}{\partial x} + 2x = 0,$$

$$\frac{\partial u}{\partial x}\cos v + u(-\sin v)\frac{\partial v}{\partial x} = 0.$$

Similarly, differentiating with respect to y, with x constant,

$$\frac{\partial u}{\partial y}\sin v + u \cos v \frac{\partial v}{\partial y} = 0,$$

$$\frac{\partial u}{\partial y}\cos v + u(-\sin v)\frac{\partial v}{\partial y} - 2y = 0.$$

Solving the last four equations for the required partial derivatives,

$$\frac{\partial u}{\partial x} = -2x \sin v, \qquad \frac{\partial v}{\partial x} = -\frac{2x}{u} \cos v,$$

$$\frac{\partial u}{\partial y} = 2y \cos v, \qquad \frac{\partial v}{\partial y} = -\frac{2y}{u} \sin v. \qquad \square$$

When the equation $F(x, y, z) = 0$ defines z as a function of x and y, the method of Problem 2.1 shows that (in the compact suffix notation)

$$F_x + F_z z_x = 0, \quad F_y + F_z z_y = 0,$$

giving $\qquad\qquad z_x = -F_x/F_z, \quad z_y = -F_y/F_z.$ \hfill (2.8)

Note that the denominator F_z does not vanish when F satisfies the conditions of the existence theorem quoted at the beginning of this section.

Again, if the simultaneous equations

$$F(x, y, u, v) = 0, \qquad G(x, y, u, v) = 0 \qquad (2.9)$$

define u and v as functions of x and y, as in Problem 2.2, then by differentiating each equation partially with respect to x and then y,

$$F_x + F_u u_x + F_v v_x = 0, \qquad F_y + F_u u_y + F_v v_y = 0, \qquad (2.10)$$

$$G_x + G_u u_x + G_v v_x = 0, \qquad G_y + G_u u_y + G_v v_y = 0. \qquad (2.11)$$

Solving,

$$u_x = - \begin{vmatrix} F_x & F_v \\ G_x & G_v \end{vmatrix} \Bigg/ \begin{vmatrix} F_u & F_v \\ G_u & G_v \end{vmatrix},$$

$$v_x = - \begin{vmatrix} F_u & F_x \\ G_u & G_x \end{vmatrix} \Bigg/ \begin{vmatrix} F_u & F_v \\ G_u & G_v \end{vmatrix},$$

$$\text{(2.12)}$$

with two similar relations in which x is replaced by y. It is naturally assumed that the denominator $J = F_u G_v - F_v G_u \neq 0$.

An existence theorem in the case of (2.9) applies when the equations are satisfied for one set of values x_0, y_0, u_0, v_0, if F, G and their first partial derivatives are continuous, and $J \neq 0$, near (x_0, y_0, u_0, v_0). The theorem states that in a certain region of the xy plane containing (x_0, y_0) there is precisely one pair of differentiable functions $u = f(x, y)$, $v = g(x, y)$ which reduce (2.9) to identities and are such that $u_0 = f(x_0, y_0)$, $v_0 = g(x_0, y_0)$.

Functional determinants of the type occurring in (2.12) are called *Jacobians*, the denominator in that equation being denoted by $\partial(F, G)/\partial(u, v)$. Similarly, if F, G, H are functions of variables $u, v, w, x_1, \ldots x_n$, then the Jacobian of F, G, H with respect to u, v, w is

22

$$\frac{\partial(F, G, H)}{\partial(u, v, w)} = \begin{vmatrix} F_u & F_v & F_w \\ G_u & G_v & G_w \\ H_u & H_v & H_w \end{vmatrix}. \tag{2.13}$$

In the latter case, if the equations $F = 0$, $G = 0$, $H = 0$ define u, v, w as functions of the x's, the procedure leading to (2.12) shows that for a typical derivative we have

$$\frac{\partial u}{\partial x_i} = -\frac{\partial(F, G, H)/\partial(x_i, v, w)}{\partial(F, G, H)/\partial(u, v, w)}, \qquad (i = 1, 2, \ldots n), \tag{2.14}$$

there being five other similar forms.

Problem 2.3 If the equation $F(x, y, z) = 0$ can be solved for any one of the variables x, y, z in terms of the other two, show that

$$\left(\frac{\partial x}{\partial y}\right)_z \left(\frac{\partial y}{\partial z}\right)_x \left(\frac{\partial z}{\partial x}\right)_y = -1,$$

the variable outside each bracket being kept constant during the differentiation.

Solution. First, regarding x as a function of y and z, we get by differentiating with respect to y,

$$F_x x_y + F_y = 0,$$

i.e.
$$(x_y)_z = -F_y/F_x. \tag{2.15}$$

In like manner we find

$$(y_z)_x = -F_z/F_y, \qquad (z_x)_y = -F_x/F_z. \tag{2.16}$$

Thus, forming the product of (2.15), (2.16),

$$(x_y)_z (y_z)_x (z_x)_y = -1. \qquad \square$$

2.2 Functional Dependence

Consider the equations

$$u = x^2 + y + 1, \qquad v = x^4 + 2x^2 y + y^2 - x^2 - y. \tag{2.17}$$

Since there is an identity between u and v, namely

$$v = (u - 1)^2 - (u - 1) = u^2 - 3u + 2,$$

we say that u and v are *functionally dependent*. Generally. if $u_1, u_2, \ldots, u_m$ are m functions of n variables $x_1, x_2, \ldots, x_n$, then the u's are said to be functionally dependent if there is an identical relation of the form $F(u_1, u_2, \ldots u_m) = 0$.

In the following problems it is assumed that F is a continuous function with continuous partial derivatives.

Problem 2.4 (i) Prove that if $u(x, y)$, $v(x, y)$ are functionally dependent,

23

in a region of the xy plane, then

$$\frac{\partial(u, v)}{\partial(x, y)} \equiv 0. \tag{2.18}$$

(ii) Verify (2.18) in the case where u and v are given by (2.17).

Solution. (i) Given that there is an identical relation of the form $F(u, v) = 0$, we get on differentiating partially with respect to x and y in turn

$$F_u u_x + F_v v_x = 0, \tag{2.19}$$
$$F_u u_y + F_v v_y = 0. \tag{2.20}$$

For consistency,

$$\begin{vmatrix} u_x & v_x \\ u_y & v_y \end{vmatrix} = \frac{\partial(u, v)}{\partial(x, y)} = 0, \tag{2.21}$$

unless both F_u and F_v vanish identically.

To deal with the latter case, note that if u and v are such that the Jacobian (2.21) is continuous and is non-zero at any one point, then it is non-zero in a neighbourhood N of this point. Therefore, by (2.19), (2.20), F_u and F_v vanish throughout N, and so F is not dependent on u or v and the equation $F = 0$ defines no functional relationship.

(ii) For $u = x^2 + y + 1$, $v = x^4 + 2x^2 y + y^2 - x^2 - y$, the Jacobian (2.21) is

$$\frac{\partial(u, v)}{\partial(x, y)} = \begin{vmatrix} 2x & 4x^3 + 4xy - 2x \\ 1 & 2x^2 + 2y - 1 \end{vmatrix}$$
$$= 2x(2x^2 + 2y - 1) - (4x^3 + 4xy - 2x) = 0. \quad \square$$

Problem 2.5 Show that the condition (2.18) is *sufficient* for $u(x, y)$, $v(x, y)$ to be functionally related.

Solution. Write

$$u = f(x, y), \qquad v = g(x, y). \tag{2.22}$$

Neglecting the trivial case where all elements of the determinant (2.18) are zero, we may suppose that $u_x \neq 0$ (otherwise we interchange the notation u and v, or x and y). Therefore the first of (2.22) in the form $f(x, y) - u = 0$ determines x as a function of y and u, according to the existence theorem on page 20. Substitution of this function for x in the second of (2.22) gives rise to a relation of the form

$$v = G(u, y). \tag{2.23}$$

If we can show that G_y is identically zero, it will follow from (2.23) that u and v are functionally related. Now, by (2.23),

$$v_x = G_u u_x, \qquad v_y = G_u u_y + G_y,$$

and so the given condition (2.18) gives

$$0 = \begin{vmatrix} u_x & u_y \\ v_x & v_y \end{vmatrix} = \begin{vmatrix} u_x & u_y \\ G_u u_x & G_u u_x + G_y \end{vmatrix} = u_x G_y,$$

on expanding the determinant. But $u_x \neq 0$, and therefore $G_y = 0$. The result follows. □

Problem 2.6 Show by the Jacobian test that there is a functional relationship between

$$u = 2 \ln x + \ln y, \quad v = e^{x\sqrt{y}}, \quad (x, y > 0), \tag{2.24}$$

and determine it.

Solution. We have

$$\begin{vmatrix} u_x & u_y \\ v_x & x_y \end{vmatrix} = \begin{vmatrix} 2/x & 1/y \\ \sqrt{y}e^{x\sqrt{y}} & e^{x\sqrt{y}}x/2\sqrt{y} \end{vmatrix} = 0,$$

and therefore there is a functional relationship between u and v.

Solving the first of (2.24) for y, we find $y = e^u/x^2$. Substituting for y in the second of (2.24),

$$\ln v = x\sqrt{y} = e^{\frac{1}{2}u},$$

i.e. $$e^{\frac{1}{2}u} - \ln v = 0.$$

This result may also be found by inspection. □

Problem 2.7 (i) Show that if $u(x, y, z)$, $v(x, y, z)$, $w(x, y, z)$ are functionally related, then

$$\begin{vmatrix} u_x & u_y & u_z \\ v_x & v_y & v_z \\ w_x & w_y & w_z \end{vmatrix} = 0. \tag{2.25}$$

(ii) Assuming that the converse result is also valid, show that there is a functional relationship between

$$u = y^2(y - z + x) - x^2(x + y - z),$$
$$v = x + y,$$
$$w = y^2 - x^2 - yz + xz.$$

Solution. (i) Let the relation be $F(u, v, w) = 0$. Differentiation with respect to x, y and z, in turn, gives

$$F_u u_x + F_v v_x + F_w w_x = 0,$$

with two further equations in which x is replaced by y and by z. Now, unless F_u, F_v, F_w all vanish (in which case $F = 0$ does not define a functional relationship) we must have for consistency of the last three equations the determinantal condition (2.25).

25

(ii) For the given functions, we find on forming the necessary derivatives

$$
\begin{vmatrix} u_x & u_y & u_z \\ v_x & v_y & v_z \\ w_x & w_y & w_z \end{vmatrix} = \begin{vmatrix} y^2 - 3x^2 - 2xy + 2xz & 3y^2 - 2yz + 2xy - x^2 & x^2 - y^2 \\ 1 & 1 & 0 \\ z - 2x & 2y - z & x - y \end{vmatrix}
$$

$$
= 0, \tag{2.26}
$$

as we find (for example) on expanding on the middle row and reducing. Hence there is a functional relationship between u, v and w. It is, in fact, $u = vw$. $\qquad\square$

As an extension of the previous results, suppose that we are given m functions of n variables, $u_1(x_1, \ldots, x_n), \ldots, u_m(x_1, \ldots, x_n)$. If A denotes the $m \times n$ matrix whose element in the ith row and jth column is $\partial u_i / \partial x_j$, then there is a functional relationship of the form $F(u_1, \ldots, u_m) = 0$ if the rank of A is less than m. (The *rank* of a matrix is the maximum number of linearly independent rows or columns.) When the rank is r ($<m$), there are exactly $m - r$ independent relations between the u's.

For example, the 3×3 matrix (2.26) has rank less than 3 since the determinant is zero. But, by inspection, the last two rows are linearly independent, (as they are not proportional), and so the rank is at least 2. Thus the rank is exactly 2 and there is only 1 ($= 3 - 2$) relations between u, v, and w, namely $u = vw$.

Problem 2.8 How many independent functional relations are there between the functions

$$
t = 2x + y, \qquad\qquad u = x + z,
$$
$$
v = 2x^2 + xy + yz + 2xz, \qquad w = x + y - z?
$$

Solution. We need to consider the rank of the 4×3 matrix

$$
\begin{pmatrix} t_x & t_y & t_z \\ u_x & u_y & u_z \\ v_x & v_y & v_z \\ w_x & w_y & w_z \end{pmatrix} = \begin{pmatrix} 2 & 1 & 0 \\ 1 & 0 & 1 \\ 4x + y + 2z & x + z & 2x \\ 1 & 1 & -1 \end{pmatrix}. \tag{2.27}
$$

The rank cannot exceed the number of columns, 3, and since $m = 4$ there is at least one functional relation between t, u, v and w. (This is automatically the case as there are more functions than variables.) Since the first two rows of the matrix are evidently linearly independent, the rank is at least 2. We must determine whether the rank is 2 or 3.

An alternative definition of rank is useful here: if a matrix contains at least one p-rowed square submatrix with non-vanishing determinant, while every $(p+1)$-rowed square submatrix has determinant zero, then

26

the rank is p. The four 3-rowed submatrices in (2.27), formed by deleting any one row, are all found to have determinant zero and so we conclude that the rank is less than 3. It is therefore 2.

It follows that the number of independent functional relationships is $4-2 = 2$. These may be expressed in various ways; one way is $v = tu$, $w = t-u$. □

2.3 Properties of Jacobians

We assume in this section that all relevant derivatives exist and are continuous.

Problem 2.9 If $u = u(x, y)$, $v = v(x, y)$, where x and y are functions of independent variables r and s, show that

$$\frac{\partial(u, v)}{\partial(r, s)} = \frac{\partial(u, v)}{\partial(x, y)} \frac{\partial(x, y)}{\partial(r, s)}. \tag{2.28}$$

Solution. We note the similarity between this equation and the rule for differentiating a function of a function in the calculus of one variable. The proof of the result is simple, since

$$\text{r.h.s.} = \begin{vmatrix} u_x & u_y \\ v_x & v_y \end{vmatrix} \begin{vmatrix} x_r & x_s \\ y_r & y_s \end{vmatrix}$$

$$= \begin{vmatrix} u_x x_r + u_y y_r & u_x x_s + u_y y_s \\ v_x x_r + v_y y_r & v_x x_s + v_y y_s \end{vmatrix} = \begin{vmatrix} u_r & u_s \\ v_r & v_s \end{vmatrix} = \text{l.h.s.}$$

where we have used the matrix multiplication rule for multiplying determinants. □

Problem 2.10 If $u = u(x, y)$, $v = v(x, y)$ can be solved to give inverse relations $x = x(u, v)$, $y = y(u, v)$, show that

$$\frac{\partial(u, v)}{\partial(x, y)} = 1 \bigg/ \frac{\partial(x, y)}{\partial(u, v)}. \tag{2.29}$$

Solution. Note that neither Jacobian in (2.29) can vanish, since either pair of equations relating u, v with x, y is invertible.

By (2.28), with r, s replaced by u, v (which have roles as both independent and dependent variables), since

$$\frac{\partial(u, v)}{\partial(u, v)} = \begin{vmatrix} 1 & 0 \\ 0 & 1 \end{vmatrix} = 1,$$

we have

$$1 = \frac{\partial(u, v)}{\partial(x, y)} \frac{\partial(x, y)}{\partial(u, v)},$$

whence (2.29) follows immediately. □

Problem 2.11 If u, v, w are functions of independent variables $x_1, x_2,$ $x_3, x_4,$ and

$$A = \frac{\partial(u, v, w)}{\partial(x_2, x_3, x_4)}, \qquad B = \frac{\partial(u, v, w)}{\partial(x_3, x_4, x_1)},$$

$$C = \frac{\partial(u, v, w)}{\partial(x_4, x_1, x_2)}, \qquad D = \frac{\partial(u, v, w)}{\partial(x_1, x_2, x_3)},$$

show that
$$\frac{\partial A}{\partial x_1} - \frac{\partial B}{\partial x_2} + \frac{\partial C}{\partial x_3} - \frac{\partial D}{\partial x_4} = 0. \tag{2.30}$$

Solution.   Let $\partial_1 = \partial/\partial x_1$, $u_1 = \partial u/\partial x_1$, etc. Consider

$$\frac{\partial A}{\partial x_1} = \partial_1 \begin{vmatrix} u_2 & u_3 & u_4 \\ v_2 & v_3 & v_4 \\ w_2 & w_3 & w_4 \end{vmatrix}. \tag{2.31}$$

By a well-known procedure for differentiating determinants, this derivative is the sum of the three determinants obtained from A by differentiating the elements of one row while leaving the other two rows unchanged. On expanding each of these three determinants on its differentiated row, we obtain an expressiion for $\partial A/\partial x_1$ as a sum of 2×2 minors of A multiplied by a second (mixed) derivative of u. For example, the first term in the sum is

$$u_{12} \begin{vmatrix} v_3 & v_4 \\ w_3 & w_4 \end{vmatrix}. \tag{2.32}$$

Now consider the full expression (2.30); this can be written as a formal determinant

$$A_1 - B_2 + C_3 - D_4 = \begin{vmatrix} \partial_1 & \partial_2 & \partial_3 & \partial_4 \\ u_1 & u_2 & u_3 & u_4 \\ v_1 & v_2 & v_3 & v_4 \\ w_1 & w_2 & w_3 & w_4 \end{vmatrix}, \tag{2.33}$$

where we expand on the first row and always keep a differential operator to the left of the function on which it acts. By dealing with the remaining three terms in this expansion in the same way as (2.31), we can express (2.33) as a finite series of terms like (2.32), in which the two indices appearing inside the minor (e.g. 3, 4) are distinct from those of the second derivative outside the minor, (1, 2).

Inspection of (2.33) shows that the only terms involving u_{12} (or u_{21}) in the series are

$$u_{12} \begin{vmatrix} v_3 & v_4 \\ w_3 & w_4 \end{vmatrix}, \qquad -u_{21} \begin{vmatrix} v_3 & v_4 \\ w_3 & w_4 \end{vmatrix},$$

whose sum is zero. By cyclic permutation of indices and of the variables u, v, w, it follows that the coefficient of every second derivative of u, v or w in the series expansion of (2.33) likewise vanishes. Hence the value of the determinant (2.33) is zero, as was to be proved. $\square$

2.4 Transformations So far, a rule which associates with each member of one set, a particular member of a second set, has been termed a *function*. It is also called a *mapping*.

In many applications, where changes of variable are concerned, we have to deal with formulae of the form

$$x = f(u, v), \qquad y = g(u, v), \tag{2.34}$$

which map points in some region of the uv plane onto points of a corresponding region of the xy plane. Such mappings between sets of points are often called *transformations*. We may also consider transformations like

$$x = f(u, v, w), \quad y = g(u, v, w), \quad z = h(u, v, w),$$

where points in three dimensions are involved. (We shall not need to consider here another possibility; that the number of variables $u, v, \ldots$ is different from the number of variables $x, y, \ldots$.)

In (2.34), the point (x, y) corresponding to a particular choice of u and v forms the *image* of the point (u, v) under the transformation. A locus, such as a curve, in the uv plane will have an *image locus* in the xy plane.

If f and g are continuous with continuous partial derivatives, and $\partial(x, y)/\partial(u, v) \neq 0$, we can in principle solve (2.34) to obtain the *inverse transformation* $u = F(x, y), v = G(x, y)$. The correspondence is then *one-to-one*, each point in the appropriate region of one plane having a unique image point in the corresponding region of the other plane.

Problem 2.12 If R is the triangular region in the xy plane bounded by the lines $x + y = 2$, $y - x = -1$, $x = 0$, find the image region S in the uv plane under the transformation $x = u + 2v$, $y = v - 2u$. Find also $\partial(u, v)/\partial(x, y)$.

Solution. The line $x + y = 2$ is mapped onto the locus $(u + 2v) + (v - 2u) = 2$, i.e. $3v - u = 2$.

The line $y - x = -1$ is mapped onto $(v - 2u) - (u + 2v) = -1$, i.e. $3u + v = 1$, and the line $x = 0$ onto $u + 2v = 0$. Hence the region S which is bounded by these image lines in the uv plane is as shown in Fig. 2.1.

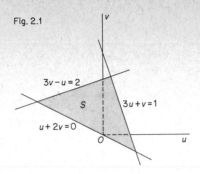

Fig. 2.1

$3v - u = 2$

S

$3u + v = 1$

$u + 2v = 0$

O

From the given relations $x = u + 2v$, $y = v - 2u$, we find

$$\frac{\partial(x, y)}{\partial(u, v)} = \begin{vmatrix} x_u & x_v \\ y_u & y_v \end{vmatrix} = \begin{vmatrix} 1 & 2 \\ -2 & 1 \end{vmatrix} = 5,$$

and therefore

$$\frac{\partial(u, v)}{\partial(x, y)} = 1 \bigg/ \frac{\partial(x, y)}{\partial(u, v)} = \frac{1}{5}.$$

Sometimes a transformation is valid and invertible except at certain points of a region, known as *singular* points, where the correspondence is not one-to-one. This occurs in many practical cases, and usually shows up in the nature of the Jacobian, which may vanish or become infinite at a singular point. □

Problem 2.13 (i) Find the region R of the xy plane which is mapped into the uv plane by the transformation

$$x = u \cosh v, \qquad y = u \sinh v. \tag{2.35}$$

(ii) Describe the curves in R whose images are the straight lines $u = u_0$ and $v = v_0$.

(iii) Indicate on a diagram the region R_1 in the xy plane whose image is the rectangular region S_1 bounded by the lines $u = 1$, $u = 2$, $v = \frac{1}{2}$, $v = 1$.

Solution. (i) Eliminating v and u in turn between (2.35) gives

$$u^2 = x^2 - y^2, \qquad \tanh v = y/x, \tag{2.36}$$

provided that $x \neq 0$. When $x = 0$, (2.35) yields $u = 0$ (since $\cosh v \neq 0$), and hence $y = 0$, so that the whole line $u = 0$ corresponds to the single point $x = 0$, $y = 0$, which is thus a singular point of the transformation.

When $x \neq 0$ we have, by (2.36), $|y/x| = |\tanh v| < 1$, and so $|y| < |x|$. Subject to this last condition, the transformation (2.35) defines a unique value of v in $-\infty < v < \infty$ for each pair of values (x, y). Again taking $|y| < |x|$, we have $u^2 > 0$ in (2.36), ensuring that u is real. By (2.35), it

30

follows that u has the same sign as x, and (2.36) then shows that u may take all real values.

Hence the region R is given by $|y| < |x|$ (shown shaded in Fig. 2.2a) together with $x = 0$, $y = 0$ as singular point. The image region is the whole of the uv plane.

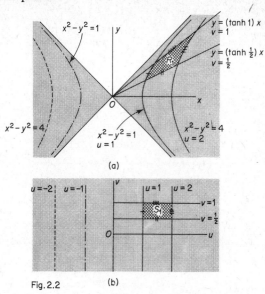

Fig. 2.2 (b)

(ii) The line $u = u_0$ is the image of a branch of the rectangular hyperbola $x^2 - y^2 = u_0^2$, the branch being such that x has the same sign as u_0. (We exclude the singular case $u_0 = 0$, dealt with in (i).) The line $v = v_0$ is the image of the line $y = (\tanh v_0)x$. Some cases are shown in Fig. 2.2.

(iii) From (ii), the branches for positive x of the hyperbolae $x^2 - y^2 = 1$ and $x^2 - y^2 = 4$ correspond respectively to $u = 1$, $u = 2$. The lines $y = (\tanh \frac{1}{2})x$ and $y = (\tanh 1)x$ correspond to $v = \frac{1}{2}$ and $v = 1$. The region R_1 is therefore as shown in the diagram.

The Jacobian $\partial(x, y)/\partial(u, v)$ corresponding to (2.35) is easily evaluated; it is found to be equal to u. We note that it vanishes at the singular point $x = 0$, $y = 0$. □

When a region R in the xy plane corresponds to a region S in the uv plane, in an invertible transformation, the boundaries of the two regions also correspond. Suppose that a point P describes the boundary of R in a definite sense (e.g. anticlockwise). Then its image point Q will describe the boundary of S, in either the same or the opposite sense. An important result is that if the Jacobian is positive the sense of description will be

31

the same for the two boundaries, and if the Jacobian is negative the senses
will be opposite.

For example, in Fig. 2.2, corresponding parts of the boundaries of R_1
and S_1 are marked with equal numbers of notches. The number of notches
increases from 1 to 4 when each boundary is traversed in the same (anticlockwise) direction, and the Jacobian u is positive.

As a second example, consider the transformation derived from (2.35)
by replacing x by $-x$. The region R_1' in the xy plane, corresponding to
S_1, is obtained by reflecting R_1 in the line $x = 0$ (Fig. 2.3). This reflection

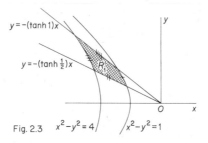

Fig. 2.3

inverts the sense in which the boundary must be traversed in order to
encounter an increasing number of notches, from 1 to 4. But we now have
$\partial(x, y)/\partial(u, v) = -u$, which is negative for $u > 0$.

Problem 2.14   Consider the transformation

$$x = \frac{\sinh u \cos \theta}{\cosh u - \cos v}, \quad y = \frac{\sinh u \sin \theta}{\cosh u - \cos v}, \quad z = \frac{\sin v}{\cosh u - \cos v}, \quad (2.37)$$

where $0 \leqslant u < \infty, \quad 0 \leqslant v < 2\pi, \quad 0 \leqslant \theta < 2\pi.$

Find the region R of xyz space which maps onto the rectangular region
in $uv\theta$ space bounded by the planes $u = \frac{1}{2}$, $u = 1$, $v = \pi/6$, $v = \pi/4$,
$\theta = 0$, $\theta = \pi/2$.

Solution. We consider first more generally the three families of surfaces
whose respective equations are $u = \text{const.}$, $v = \text{const.}$, and $\theta = \text{const.}$
Since (2.37) gives $y = x \tan \theta$, it follows that the surface $\theta = \theta_0$ is a plane
through the z axis inclined at angle θ_0 to the plane $x = 0$.

If *both* u and v are kept constant, while θ is allowed to vary, then z and
$x^2 + y^2$ remain constant according to (2.37), and so the corresponding
point in xyz space describes a circle parallel to the xy plane with centre
on the z axis. Therefore, if any member of the family $u = \text{const.}$ contains
one point of any such circle, it contains the whole circle. This means
that every member is a surface of revolution about Oz, and the same
applies to each curve of the family $v = \text{const.}$ We can find all further

32

properties of the two families by investigating their curves of intersection with the plane $\theta = 0$ (i.e. $y = 0$).

When $\theta = 0$, we get from (2.37)

$$(x - \coth u)^2 + z^2 = \operatorname{cosech}^2 u, \quad x^2 + (z - \cot v)^2 = \operatorname{cosec}^2 v. \quad (2.38)$$

The first of these equations shows that the surface $u = u_0$ meets the plane $y = 0$ in xyz space in a circle of radius $\operatorname{cosech} u_0$ centred at $x = \coth u_0$, $z = 0$. Similarly, the second of (2.38) shows that the surface $v = v_0$ meets the plane $y = 0$ in a circle of radius $\operatorname{cosec} v_0$ centred at $x = 0$, $z = \cot v_0$. (The two families of circles $u = \mathrm{const.}$, $v = \mathrm{const.}$ in the plane $y = 0$ are, in fact, orthogonal families of *coaxal* circles, with $x = \pm 1$, $z = 0$ as limiting points. It follows that each member of the first family is a torus; u, v, θ are *toroidal coordinates*.)

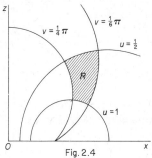

Fig. 2.4

The required region R is obtained by rotating the shaded area in Fig. 2.4 through an angle $\frac{1}{2}\pi$ about $0z$. The plane ends of the region are $y = 0$ and $x = 0$.

Note that the denominators in (2.37) vanish when $u = v = 0$, which is a line of singular points of the transformation in $uv\theta$ space. ◻

Further properties of transformations will be found in Chapter 4.

EXERCISES

1. If $x^2 + yu + zu^2 = xyz$ defines u as a function of x, y and z, find u_x, u_y, u_z.

2. If the equations

$$u^3 x - yv = u, \qquad v^3 y - xu = v,$$

define u and v as functions of x and y, find u_x and v_x.

3. Test whether there is a functional relationship between

$$u = x - y + z, \quad v = xz - (x + z)y, \quad w = x^2 + y^2 + z^2,$$

and if there is, determine it.

33

4. How many independent functional relationships are there between
$t = x-y$, $u = x+3y+2z$, $v = y(y+z)-x(x+z)$, $w = x+y+z$?

5. For the transformation $u = x+y^2$, $v = y-x^2$, evaluate the Jacobian $\partial(x, y)/\partial(u, v)$.

6. Suggest a generalization of (2.29) applying to an invertible transformation

$$u = u(x, y, z), \quad v = v(x, y, z), \quad w = w(x, y, z).$$

If $u = x+y+z$, $v = x^2+z^2$, $w = (x+y)z$, find $\partial(x, y, z)/\partial(u, v, w)$, and give the values of x, y and z for which the result is valid.

7. The rectangular region R in the xy plane bounded by the lines $x = \ln a$, $x = \ln b$, $y = \frac{1}{2}\pi$, $y = \pi$ ($b > a > 0$) is mapped onto a region S in the uv plane under the transformation

$$u = e^x\cos y, \quad v = e^x\sin y.$$

Show that the straight lines $x = x_0$ and $y = y_0$ in R map onto parts of circles and of straight lines through the origin, respectively, in the uv plane, Sketch the region S.

Evaluate $\partial(u, v)/\partial(x, y)$. If a point P moves around the boundary of R in a given sense, does the image point Q under the transformation describe the boundary of S in the same or opposite sense?

Chapter 3

Taylor's Theorem and Applications

3.1 Taylor's Theorem in Two Variables   If $f(x)$ possesses a continuous $(n+1)$th derivative in the interval $a \leqslant x \leqslant a+h$, then

$$f(a+h) = f(a) + hf'(a) + \ldots + \frac{h^n}{n!} f^{(n)}(a) + R_n, \tag{3.1}$$

where the *remainder* R_n is given by

$$R_n = \frac{h^{n+1}}{(n+1)!} f^{(n+1)}(c), \tag{3.2}$$

for some point c such that $a < c < a+h$. This result is known as *Taylor's theorem in one variable*. (The particular case where $n = 0$ is also called the *mean-value theorem*.)

There are alternative expressions for R_n. That given in (3.2) is *Lagrange's* form, while *Cauchy's* form is

$$R_n = \frac{h^{n+1} p^n}{n!} f^{(n+1)}(c), \tag{3.3}$$

where c is a point such that $a < c < a+h$ (though not necessarily the same point as in (3.2)), and $p = 1-(c-a)/h$. (Note that $0 < p < 1$.)

We can extend Taylor's theorem to functions of several variables; it will suffice here to state the theorem for two independent variables. Let $f(x, y)$ and its partial derivatives of all orders up to $n+1$ be continuous in a neighbourhood of each point on the line-segment PQ, where P is the point (a, b) and Q is the point $(a+h, b+k)$. Then

$$f(a+h,\, b+k) = f(a, b) + \left(h\frac{\partial}{\partial x} + k\frac{\partial}{\partial y} \right) f(a,b) + \ldots$$

$$+ \frac{1}{n!} \left(h\frac{\partial}{\partial x} + k\frac{\partial}{\partial y} \right)^n f(a, b) + R_n, \tag{3.4}$$

where

$$R_n = \frac{1}{(n+1)!} \left(h\frac{\partial}{\partial x} + k\frac{\partial}{\partial y} \right)^{n+1} f(c, d), \tag{3.5}$$

and

$$(c, d) = (a+\theta h, b+\theta k), \qquad 0 < \theta < 1,$$

is a point on the line-segment PQ. Here the notation

$$\left(h\frac{\partial}{\partial x} + k\frac{\partial}{\partial y} \right)^n f(a, b) \tag{3.6}$$

35

means that the operator is to be expanded by the binomial theorem and applied to $f(x, y)$ before setting $x = a$, $y = b$. For example,

$$\left(h\frac{\partial}{\partial x}+k\frac{\partial}{\partial y}\right)^2 f(a, b) = h^2 f_{xx}(a, b)+2hkf_{xy}(a, b)+k^2 f_{yy}(a, b).$$

The proof of (3.4), (3.5) is not difficult, if (3.1), (3.2) are assumed. We fix a, b, h, k and consider the function of one variable:

$$F(t) = f(a+th, b+tk), \qquad (0 \leqslant t \leqslant 1).$$

Then (3.1), (3.2) are used to express $F(t)$ in terms of $F(0)$ and the first n derivatives of $F(t)$ at $t = 0$, together with a remainder term. The result follows from the fact that a term such as (3.6) is simply $F^{(n)}(0)$, by the chain rule. Details are left to the reader. The case $n = 0$ is again referred to as the mean-value theorem.

In any of the above forms of Taylor's theorem (and obvious extensions to higher numbers of variables than two) an infinite series expansion is obtained when n is allowed to tend to infinity, provided that the given function possesses continuous derivatives of all orders. It represents the function provided R_n tends to zero. Such a series is known as a *Taylor series* for the function. Normally we would replace $a+h$ by x, and $b+k$ by y, in the Taylor series resulting from (3.4), thus expressing $f(x, y)$ as a series of products of powers of $x-a$ and $y-b$.

Problem 3.1 If $f(x, y) = x^2 y+2xy^2$, and $a = 1$, $h = 2$, $b = 2$, $k = 1$, find θ in the mean-value theorem:

$$f(a+h, b+k)-f(a, b) = \left(h\frac{\partial}{\partial x}+k\frac{\partial}{\partial y}\right) f(a+\theta h, b+\theta k), \qquad 0 < \theta < 1.$$
$$(3.7)$$

Solution. On substituting the given values, the left-hand side of (3.7) becomes

$$f(1+2, 2+1)-f(1, 2) = f(3, 3)-f(1, 2)$$
$$= (27+54)-(2+8) = 71.$$

Also

$$\left(h\frac{\partial}{\partial x}+k\frac{\partial}{\partial y}\right) f(x, y) = 2f_x(x, y)+f_y(x, y)$$

$$= 2(2xy+2y^2)+(x^2+4xy) = x^2+8xy+4y^2.$$

Putting $x = a+\theta h = 1+2\theta$, $y = b+\theta k = 2+\theta$, and equating the two sides of (3.7) we get

$$(1+2\theta)^2+8(1+2\theta)(2+\theta)+4(2+\theta)^2 = 71,$$

i.e.
$$120^2 + 30\theta - 19 = 0,$$
$$\theta = [-15 \pm \sqrt{(553)}]/12.$$

But $0 < \theta < 1$, which shows that the positive sign must be taken. We find, approximately, $\theta = 0.710$. ☐

Problem 3.2 Expand $e^{2x}\cos y$ as a Taylor series about $x = 0$, $y = 0$, as far as terms of third degree.

Solution. The question means that we are to take $a = 0$, $b = 0$ in the notation of this section. We thus obtain an approximate formula for the given function, for small values of x and y.

The Taylor series for $f(x, y)$ about $(0, 0)$ is

$$f(x, y) = f(0, 0) + xf_x + yf_y + \frac{1}{2!}(x^2 f_{xx} + 2xy f_{xy} + y^2 f_{yy}) + \dots, \qquad (3.8)$$

the derivatives all being evaluated at $(0, 0)$. At this point, $(0, 0)$, we find by straightforward calculation the following values for $f(x, y) = e^{2x}\cos y$ and its partial derivatives up to third order,

$$f = 1, \quad f_x = 2, \quad f_y = 0,$$
$$f_{xx} = 4, \quad f_{xy} = 0, \quad f_{yy} = -1,$$
$$f_{xxx} = 9, \quad f_{xxy} = 0, \quad f_{xyy} = -2, \quad f_{yyy} = 0.$$

Substitution in (3.8) gives the required Taylor series

$$e^{2x}\cos y = 1 + 2x + \frac{1}{2!}(4x^2 - y^2) + \frac{1}{3!}(8x^3 - 6xy^2) + \dots \quad ☐ \quad (3.9)$$

Problem 3.3 How may the series in (3.9) be modified so that it terminates with terms of third degree in x and y, while the equation remains exact?

Solution. Here we replace the third degree terms by the remainder R_2. With $a = b = 0$ in (3.5), and (h, k) replaced by (x, y) after the necessary differentiations have been performed, we have

$$R_2 = \frac{1}{3!}(x^3 f_{xxx} + 3x^2 y f_{xxy} + 3xy^2 f_{xyy} + y^3 f_{yyy}) \qquad (3.10)$$

(by expansion of the operator), where all derivatives are evaluated at $(\theta x, \theta y)$, and $0 < \theta < 1$.

Differentiating $f(x, y) = e^{2x}\cos y$, we get at (x, y)

$$f_{xxx} = 8e^{2x}\cos y, \qquad f_{xxy} = -4e^{2x}\sin y,$$
$$f_{xyy} = -2e^{2x}\cos y, \qquad f_{yyy} = e^{2x}\sin y,$$

Replacing (x, y) by $(\theta x, \theta y)$ and substituting in (3.10),

$$R_2 = \frac{1}{3!}[x^3(8\cos\theta y) + 3x^2 y(-4\sin\theta y) + 3xy^2(-2\cos\theta y) + y^3(\sin\theta y)]e^{2\theta x}$$

$$= \frac{1}{3!}[(8x^3 - 6xy^2)\cos\theta y - (12xy^2 - y^3)\sin\theta y]e^{2\theta x}. \qquad \square$$

Problem 3.4 Find the Taylor series for $f(x, y) = e^{x^2 \sin xy}$ about the point $(2, 0)$ as far as the terms of second degree. Hence obtain an approximate value for $f(1 \cdot 98, 0 \cdot 015)$.

Solution. At (2.0) we find

$$f = 1, \quad f_x = 0, \quad f_y = 8,$$
$$f_{xx} = 0, \quad f_{xy} = 12, \quad f_{yy} = 64.$$

Therefore the Taylor series for $f(x, y)$ about $(2, 0)$ is

$$f(x, y) = f(2, 0) + (x - 2)f_x(2, 0) + yf_y(2, 0)$$
$$+ \frac{1}{2!}[(x - 2)^2 f_{xx}(2, 0) + 2(x - 2)yf_{xy}(2, 0) + y^2 f_{yy}(2, 0)] + \ldots$$
$$= 1 + 8y + 12(x - 2)y + 32y^2 + \ldots .$$

If we replace x by $2 + \Delta x$, and y by Δy, in the last equation we get

$$f(2 + \Delta x, \Delta y) = 1 + 8\Delta y + 12\Delta x \Delta y + 32\Delta y^2 + \ldots .$$

The required approximation is obtained by setting $\Delta x = -0 \cdot 02$, $\Delta y = 0 \cdot 015$, so that, if terms of higher order than second in $\Delta x, \Delta y$ are neglected,

$$f(1 \cdot 98, 0 \cdot 015) = 1 + (8 + 12\Delta x + 32\Delta y)\Delta y$$
$$= 1 + (8 - 0 \cdot 24 + 0 \cdot 48)(0 \cdot 015) = 1 \cdot 124,$$

to three decimal places. For comparison, the exact value is $1 \cdot 1235$, to four decimal places. $\qquad \square$

It should be noted that the Taylor series represents a given function only if the remainder term R_n in (3.1) tends to zero as n tends to infinity. It is *not* sufficient that the infinite series converge. For example, the function defined by

$$f(x, y) = \begin{cases} e^{-1/(x^2 + y^2)}, & (x, y) \neq (0, 0), \\ 0, & (x, y) = (0, 0), \end{cases}$$

possesses continuous partial derivatives of all orders at $(0, 0)$, and all are found to vanish at this point. (They are calculated from first principles.) Hence the Taylor series about $(0, 0)$ for this function is identically zero, and does not represent $f(x, y)$,

However, in most practical cases, if the Taylor series converges it represents the function. The method of Problem 3.4 will then lead to a satisfactory approximation if sufficient terms are retained.

3.2 Maxima and Minima Let $f(x, y)$ be defined in a region R which

38

contains (x_0, y_0) as an interior point. Then f is said to have an *absolute maximum* at (x_0, y_0) if

$$f(x, y) \leqslant f(x_0, y_0) \qquad (3.11)$$

for every point (x, y) in R. If the inequality in (3.11) is reversed, then f has an *absolute minimum* at (x_0, y_0). An absolute maximum or minimum is called an *absolute extremum*.

Often we are more concerned with *relative* extrema, i.e. when either inequality is required to hold only for points (x, y) in a neighbourhood of (x_0, y_0). Thus, f has a *relative maximum* at (x_0, y_0) if the increment

$$\Delta f = f(x, y) - f(x_0, y_0) \leqslant 0 \qquad (3.12)$$

for all points (x, y) in R such that

$$|x - x_0| < h, \qquad |y - y_0| < h,$$

for some sufficiently small h ($h > 0$). A corresponding definition applies for a *relative minimum*, and both definitions allow direct extension to functions of more than two variables.

In the following, we shall assume that functions are continuous and possess continuous derivatives of all necessary orders.

Problem 3.5 Show that if $f(x, y)$ has a relative extremum at (x_0, y_0), then f_x and f_y both vanish at this point.

Solution. We treat the case where f has a relative maximum at (x_0, y_0); the case of a relative minimum merely involves the reversal of certain inequalities in the argument.

Consider (3.12) with y kept constant at the value y_0. We have

$$f(x, y_0) - f(x_0, y_0) \leqslant 0,$$

for $|x - x_0| < h$, where h is some positive number. Hence the function $F(x) = f(x, y_0)$ has a relative maximum (in the context of the calculus of one variable) at $x = x_0$. It follows from elementary theory that $F'(x_0) = 0$, i.e. $f_x(x_0, y_0) = 0$.

Similarly, by keeping x constant at the value x_0, we find that the function $G(y) = f(x_0, y)$ has a relative maximum at $y = y_0$, giving $G'(y_0) = f_y(x_0, y_0) = 0$. Hence the result. $\qquad \square$

Problem 3.6 Find the point on the plane $x + 2y - 3z = 4$ which is nearest the origin.

Solution. Let l denote the distance of a general point $P(x, y, z)$ from the origin, so that $l^2 = x^2 + y^2 + z^2$. If P lies on the given plane, we have by substitution for z,

$$l^2 = x^2 + y^2 + \tfrac{1}{9}(x + 2y - 4)^2.$$

It is evident from the nature of the problem that l^2 in the last equation must possess a minimum (both absolute and relative); this must occur where

$$\frac{\partial l^2}{\partial x} = 2x + \tfrac{2}{9}(x + 2y - 4) = 0, \tag{3.13}$$

$$\frac{\partial l^2}{\partial y} = 2y + \tfrac{4}{9}(x + 2y - 4) = 0, \tag{3.14}$$

i.e. $\qquad\qquad 5x + y = 2, \qquad 2x + 13y = 8,$

giving $x = \tfrac{2}{7}$, $y = \tfrac{4}{7}$.

Since this is the only solution of (3.13), (3.14), it must correspond to the required point. We then find from the equation of the given plane that $z = -6/7$ for these values of x and y, and so the point nearest the origin is given by

$$x = \tfrac{2}{7}, \quad y = \tfrac{4}{7}, \quad z = -\tfrac{6}{7}. \qquad\qquad \square$$

A *critical point* of the function $f(x, y)$ is a point (x_0, y_0) at which $f_x = f_y = 0$. At such a point the tangent plane to the surface $z = f(x, y)$ is horizontal. The following problem shows that a critical point need not correspond to a relative extremum.

Problem 3.7 Find any critical points of the function $f(x, y) = y^2 - x^2$, and show that the function has no relative extrema.

Solution. The equations $f_x = 0$, $f_y = 0$ give $x = y = 0$ as the only critical point. Now, along the line $y = 0$, we have

$$\Delta f = f(x, 0) - f(0, 0) = -x^2 \leqslant 0,$$

while along the line $x = 0$,

$$\Delta f = f(0, y) - f(0, 0) = y^2 \geqslant 0.$$

Hence the sign of Δf depends on the direction of the point (x, y) from the point $(0, 0)$, and so cannot correspond either to a relative maximum or a relative minimum. As there are no other critical points the result follows.

We note that the function $f(x, 0)$, as a function of one variable, has a maximum at $x = 0$, while the function $f(0, y)$ has a minimum at $y = 0$. Part of the surface $z = f(x, y)$ is shown in Fig. 3.1. $\qquad\qquad \square$

When the sign of Δf in (3.12) depends on the direction of the point $Q(x, y)$ from a critical point $P(x_0, y_0)$, as in the last problem, the critical point is called a *saddle point*. Saddle points can be more complicated than the type in Fig. 3.1. For example, the behaviour of f along PQ may alternate many times as PQ makes a complete turn about P.

Problem 3.8 Let $P(x_0, y_0)$ be a critical point of $f(x, y)$, and let r, s, t denote the values of f_{xx}, f_{xy}, f_{yy}, respectively, at P.

40

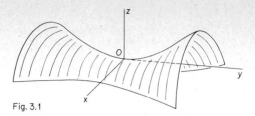

Fig. 3.1

(i) Show that a sufficient condition for P to be a relative extremum point is that

$$D \equiv rt - s^2 > 0. \tag{3.15}$$

(ii) If $D > 0$, show that the extremum is a maximum provided $r < 0$ (or $t < 0$), and is a minimum provided $r > 0$ (or $t > 0$).

(iii) If $D < 0$, show that the extremum is a saddle point.

Solution. (i) Since $f_x = f_y = 0$ at the critical point P, the Taylor expansion of $f(x, y)$ about P gives, with $x - x_0 = l, \ y - y_0 = m$,

$$\Delta f = f(x, y) - f(x_0, y_0) = \tfrac{1}{2}(l^2 f_{xx} + 2lm f_{xy} + m^2 f_{yy}), \tag{3.16}$$

where the right-hand side is evaluated at $x = x_0 + \theta l, \ y = y_0 + \theta m$, and $0 < \theta < 1$. We shall show that $\Delta f \neq 0$ for all sufficiently small l and m (not both zero), provided that $D > 0$. This means that Δf cannot change sign near P (assuming the second derivatives of f to be continuous), and we shall have a relative maximum at P if Δf is negative and a relative minimum if Δf is positive.

In fact, if Δf were to vanish and if $m \neq 0$, we get from (3.16), on dividing by $\tfrac{1}{2}m^2$, that the quadratic equation

$$(l/m)^2 f_{xx} + 2(l/m) f_{xy} + f_{yy} = 0 \tag{3.17}$$

has real roots for l/m, whence

$$f_{xx} f_{yy} - (f_{xy})^2 \leqslant 0. \tag{3.18}$$

In the case $m = 0$, we can divide (3.16) by $\tfrac{1}{2}l^2$ instead in order to obtain (3.18). But the inequality (3.18) cannot hold near P since $rt - s^2 > 0$ and the left-hand side of (3.18) is continuous. Therefore Δf does not vanish near P, which must be a relative extremum.

(ii) Since $rt - s^2 > 0$, r and t do not vanish, and have a common sign. By putting $m = 0, l \neq 0$ in (3.16) we see that the sign of Δf is the same as that of f_{xx}, and by continuity this is the sign of r. Therefore $r < 0$ (or $t < 0$) corresponds to a relative maximum, and $r > 0$ (or $t > 0$) to a relative minimum.

(iii) If $D < 0$, (3.18) holds near P. Therefore (3.17) has real roots for l/m. (If $m = 0$, the corresponding quadratic equation in m/l is considered.) Thus, the left-hand side of (3.17) changes sign as the ratio l/m is varied so

41

as to pass through a root of the equation. By (3.16), Δf likewise changes sign, which proves that P is a saddle point. (Note that l/m determines the direction of $Q(x, y)$ from $P(x_0, y_0)$.)

When $D = 0$, a further test is necessary to determine the nature of a critical point. □

Problem 3.9 Find and identify all critical points of the function
$$f(x, y) = y^3 + 3x^2 y - 3x^2 - 3y^2 + 2.$$
Solution. On putting $f_x = f_y = 0$, we find
$$x(y-1) = 0, \qquad y(y-2) + x^2 = 0.$$
The first of these equations gives $x = 0$ or $y = 1$, and using the second we find that the critical points are
$$(0, 0), \quad (0, 2), \quad (1, 1), \quad (-1, 1).$$
Now,
$$f_{xx} = 6(y-1), \quad f_{xy} = 6x, \quad f_{yy} = 6(y-1),$$
giving
$$f_{xx} f_{yy} - (f_{xy})^2 = 36[(y-1)^2 - x^2].$$
Therefore:

at $(0, 0)$, $D = 36 > 0$, $r = -6 < 0$, and f has a rel. max.,

at $(0, 2)$, $D = 36 > 0$, $r = 6 > 0$, and f has a rel. min.,

at $(1, 1)$, $D = -36 < 0$, and f has a saddle point,

at $(-1, 1)$, $D = -36 < 0$, and f has a saddle point. □

Problem 3.10 It is intended to construct an open metal water tank with a right-angled triangle as base, the sides being vertical. If the volume of the tank is to be 2 m³, what is the least area of metal that can be used?

Solution. Let the perpendicular sides of the triangular base by x m and y m respectively, and let the height be z m. Then the total surface area is
$$S = \tfrac{1}{2}xy + xz + yz + (x^2 + y^2)^{\frac{1}{2}}z. \tag{3.19}$$
But the volume is to be 2, and so
$$2 = \tfrac{1}{2}xyz, \qquad \text{i.e.} \quad z = 4/xy.$$
Substituting in (3.19) for z,
$$S = \tfrac{1}{2}xy + 4[x + y + (x^2 + y^2)^{\frac{1}{2}}]/xy. \tag{3.20}$$
For an extremum, the last equation gives
$$\frac{\partial S}{\partial x} = \tfrac{1}{2}y - \frac{4}{x^2}\left[1 + \frac{y}{(x^2 + y^2)^{\frac{1}{2}}}\right] = 0, \tag{3.21}$$

42

$$\frac{\partial S}{\partial y} = \tfrac{1}{2}x - \frac{4}{y^2}\left[1 + \frac{x}{(x^2+y^2)^{\frac{3}{2}}}\right] = 0. \tag{3.22}$$

Multiplying (3.21) by x^2, (3.22) by y^2, and subtracting,
$$(x-y)\left[\tfrac{1}{2}xy + 4(x^2+y^2)^{-\frac{1}{2}}\right] = 0.$$

The expression in square brackets cannot vanish since x and y are positive, and so we must have $x = y$. Substituting y for x in (3.21) gives
$$y^3 = 8(1 + \tfrac{1}{2}\sqrt{2}),$$
so that $\qquad y = x = 2(1 + \tfrac{1}{2}\sqrt{2})^{\frac{1}{3}} = 2{\cdot}39 \quad$ (approx.).

Thus $\qquad z = 4/xy = (1 + \tfrac{1}{2}\sqrt{2})^{-\frac{2}{3}} = 0{\cdot}700 \quad$ (approx.),

and the minimum surface area of the tank is, by (3.20),
$$S = 6(1 + \tfrac{1}{2}\sqrt{2})^{\frac{2}{3}} = 8{\cdot}57\,\mathrm{m}^2,$$
approximately. This is the least area of metal required.

Here we have deduced from the nature of the problem that the only critical point of S, as a function of x and y, must be a minimum. We can verify this by the method used in the last two problems. In fact, when $y = x$ we easily find
$$S_{xx} = S_{yy} = (8 + 5\sqrt{2})/x^3, \qquad S_{xy} = (4 + \sqrt{2})/x^3,$$
so that
$$S_{xx}S_{yy} - (S_{xy})^2 > 0, \qquad S_{xx} > 0,$$
which proves that a critical point for which $x = y$ is certainly a minimum. □

When D vanishes at a critical point, the test used above is inapplicable. One commonly occurring case is that $r = s = t = 0$. The sign of Δf then depends on the third derivative terms in the Taylor expansion of $f(x, y)$ about the critical point (x_0, y_0). A detailed examination of these terms shows that there is no relative extremum unless they all vanish, in which case it is necessary to investigate the terms involving fourth derivatives to complete the test.

Problem 3.11 Verify that $f(x, y) = x^2y^2(x+y+1) - 1$ has a critical point at $(0, 0)$, and determine its nature.

Solution. At $(0, 0)$, we have
$$f_x = 2xy^2(x+y+1) + x^2y^2 = 0,$$
$$f_y = 2x^2y(x+y+1) + x^2y^2 = 0,$$
and so $P(0, 0)$ is a critical point.

Further differentiation shows that all second order derivatives vanish at this point, and so $D = 0$. Third order derivatives are also found to

vanish at $(0, 0)$, as do those of fourth order except for
$$\partial^4 f/\partial x^2\, \partial y^2 = 12(x+y)+4 = 4.$$
The Taylor expansion of $f(x, y)$ is found to be
$$\Delta f = f(x, y) - f(0, 0) = x^2 y^2 [1 + \tfrac{5}{4}(\theta x + \theta y)],$$
where $0 < \theta < 1$, the expression in square brackets being obtained by evaluating the fourth derivatives of f at $(\theta x, \theta y)$. Since Δf is positive for all small x and y, f has a relative minimum at $(0, 0)$. (This result may otherwise be obtained by inspection of f.) □

A function $f(x, y, z)$ of three variables has a relative maximum at $P(x_0, y_0, z_0)$ if
$$\Delta f = f(x, y, z) - f(x_0, y_0, z_0) \leqslant 0$$
whenever $|x - x_0| < h, |y - y_0| < h, |z - z_0| < h$, for sufficiently small h. A corresponding definition applies for a relative minimum. A necessary condition is that P be a critical point, i.e. that all the first partial derivatives of f vanish there.

A generalization of the D test is that, for a minimum it is sufficient that

$$f_{xx} > 0, \qquad \begin{vmatrix} f_{xx} & f_{xy} \\ f_{yx} & f_{yy} \end{vmatrix} > 0, \qquad \begin{vmatrix} f_{xx} & f_{xy} & f_{xz} \\ f_{yx} & f_{yy} & f_{yz} \\ f_{zx} & f_{zy} & f_{zz} \end{vmatrix} > 0, \qquad (3.23)$$

whereas if the signs in these inequalities alternate, starting with $f_{xx} < 0$, then the critical point is a maximum. The results extend to any number of variables.

Problem 3.12 Find and identify any critical points of
$$f(x, y, z) = e^{x+y+z}/(1+e^x)(e^x+e^y)(e^y+e^z)(1+e^z).$$

Solution. Differentiating with respect to x, regarding f as a product of five factors, we obtain
$$f_x = f - \frac{e^x}{1+e^x}f - \frac{e^x}{e^x+e^y}f,$$
which vanishes (since $f \neq 0$) only when
$$(1+e^x)(e^x+e^y) = e^x[(e^x+e^y)+1+e^x],$$
i.e. $e^{2x} = e^y$, giving $y = 2x$.

By symmetry, $f_z = 0$ when $y = 2z$, and a similar calculation gives
$$f_y = f - \frac{e^y}{e^x+e^y}f - \frac{e^y}{e^y+e^z}f,$$
whence we find $f_y = 0$ when $2y = x + z$.

44

The critical points are therefore determined by the simultaneous equations

$$y = 2x, \quad y = 2z, \quad 2y = x+z,$$

which possess only the solution $x = y = z = 0$.

At this point, a straightforward calculation yields

$$f_{xx} = f_{yy} = f_{zz} = -\tfrac{1}{32},$$
$$f_{xy} = f_{yz} = \tfrac{1}{64}, \qquad f_{xz} = 0.$$

Applying the test (3.23), we find

$$f_{xx} < 0, \quad \begin{vmatrix} f_{xx} & f_{xy} \\ f_{yx} & f_{yy} \end{vmatrix} = (-\tfrac{1}{32})(-\tfrac{1}{32}) - (\tfrac{1}{64})^2 > 0,$$

while the 3×3 determinant in (3.23) is found to have the value (on taking out a factor $\tfrac{1}{64}$ from each element)

$$\frac{1}{(64)^3} \begin{vmatrix} -2 & 1 & 0 \\ 1 & -2 & 1 \\ 0 & 1 & -2 \end{vmatrix} = \frac{-4}{(64)^3} < 0.$$

Hence f has a relative maximum at $(0, 0, 0)$. $\qquad\square$

3.3 Constraints: Undetermined Multipliers Suppose that we require to find extrema of a function $f(x, y, z)$, where x, y and z are not independent variables but are subject to a condition or *constraint* $g(x, y, z) = 0$. In some cases we can use the constraint equation to eliminate one of the variables from f, as in Problems 3.6 and 3.10. But in others the elimination is impracticable, and it is convenient to use the method of an *undetermined multiplier*.

This method uses the fact that if f has an extremum at a particular point under the given constraint, then so does the function $\phi = f + \lambda g$, where λ is an arbitrary constant, since the additional term is zero. If we regard x and y as independent variables, with z determined by the equation $g = 0$, then for an extremum of $\phi(x, y, z)$ we need

$$\phi_x + \phi_z z_x = 0, \qquad \phi_y + \phi_z z_y = 0. \tag{3.24}$$

Now choose λ so that $\phi_z = 0$ at the extremum, whence by (3.24) we get $\phi_x = 0, \phi_y = 0$, i.e.

$$f_x + \lambda g_x = 0, \quad f_y + \lambda g_y = 0, \quad f_z + \lambda g_z = 0. \tag{3.25}$$

In solving problems we write down these equations together with

$$g = 0, \tag{3.26}$$

the four relations (3.25), (3.26) determining x, y, z, λ. In most cases the value of λ is not of interest; it may then be left *undetermined*. We note that the choice of λ to satisfy $\phi_z = 0$ requires only $g_z \neq 0$; if this is not the case, the above results may be obtained by singling out x or y instead of z as dependent variable.

Problem 3.13   A rectangular box is just contained within the ellipsoid $2x^2 + 3y^2 + z^2 = 18$, and each edge is parallel to one of the coordinate axes. Find its greatest possible volume.

Solution. The ellipsoid is symmetrical about each of the planes $x = 0$, $y = 0$, $z = 0$, its principal axes being segments of the coordinate axes. By symmetry, it follows that the vertices of the box all lie on the ellipsoid and their coordinates will be of the form $(\pm x, \pm y, \pm z)$, where

$$g(x, y, z) \equiv 2x^2 + 3y^2 + z^2 - 18 = 0. \tag{3.27}$$

The volume of the box is $V = 8xyz$, and we require the maximum value of V subject to the constraint (3.27). The equations corresponding to (3.25), with V in place of f, are

$$8yz + 4\lambda x = 0, \tag{3.28}$$
$$8xz + 6\lambda y = 0, \tag{3.29}$$
$$8xy + 2\lambda z = 0. \tag{3.30}$$

Multiplying the last three equations respectively by x, y and z, and subtracting the second from the first, and the third from the second, we get

$$\lambda(4x^2 - 6y^2) = 0, \qquad \lambda(6y^2 - 2z^2) = 0. \tag{3.31}$$

Since $y > 0$, $z > 0$, we have $\lambda \neq 0$, by (3.28). Hence by (3.31),

$$z^2 = 3y^2 = 2x^2. \tag{3.32}$$

Substituting for x and y from (3.32) in (3.27), we get $z^2 = 6$, and so

$$x = \sqrt{3}, \quad y = \sqrt{2}, \quad z = \sqrt{6}, \tag{3.33}$$

giving
$$V = 8xyz = 48.$$

As there is only one critical point for positive x, y and z, it follows that the last equation gives the maximum possible volume of the box.

For interest, we can also determine λ. By (3.28), (3.33), we find $\lambda = -4$. □

The method of multipliers applies in modified form to the problem of finding critical points of a function of any number of variables, when these are subject to any lesser number of constraints. If, as before, there are three variables x, y, z, but these are subject to two independent constraints of the form $g(x, y, z) = 0$, $h(x, y, z) = 0$, then the equations to determine the critical points of $f(x, y, z)$ are

$$f_x + \lambda g_x + \mu h_x = 0, \tag{3.34}$$

$$f_y + \lambda g_y + \mu h_y = 0, \tag{3.35}$$
$$f_z + \lambda g_z + \mu h_z = 0, \tag{3.36}$$
$$g = 0, \qquad h = 0. \tag{3.37}$$

Here, there are two undetermined multipliers, λ and μ. The number is always the same as the number of constraints.

Problem 3.14 Find the shortest distance from the origin to the line of intersection of the two planes $lx + my + nz = p$, $l'x + m'y + n'z = p'$, where $l^2 + m^2 + n^2 = 1$, $l'^2 + m'^2 + n'^2 = 1$.

Solution. We have to minimize the expression $r^2 = x^2 + y^2 + z^2$, the square of the distance of the point (x, y, z) from the origin, under the constraint equations of the two given planes. Hence, corresponding to (3.34), ..., (3.37) we have

$$2x + \lambda l + \mu l' = 0, \tag{3.38}$$
$$2y + \lambda m + \mu m' = 0, \tag{3.39}$$
$$2z + \lambda n + \mu n' = 0, \tag{3.40}$$
$$lx + my + nz - p = 0, \qquad l'x + m'y + n'z - p' = 0. \tag{3.41}$$

These equations determine x, y, z at a critical point, together with λ and μ. Clearly the critical point must correspond to a minimum for r^2.

Multiplying (3.38), (3.39), (3.40) by x, y, z, respectively, and adding,

$$2r^2 + \lambda p + \mu p' = 0.$$

Multiplying the same equations by l, m, n and adding,

$$2p + \lambda + \mu k = 0,$$

where $k = ll' + mm' + nn'$. Finally, multiplying the equations by l', m', n', respectively, and adding,

$$2p' + \lambda k + \mu = 0.$$

Eliminating λ and μ from the last three equations,

$$\begin{vmatrix} 2r^2 & p & p' \\ 2p & 1 & k \\ 2p' & k & 1 \end{vmatrix} = 0,$$

giving on expansion,

$$r^2(1 - k^2) - p(p - kp') + p'(pk - p') = 0.$$

Solving for r, we obtain for the required minimum distance

$$r = [(p^2 + p'^2 - 2kpp')/(1 - k^2)]^{\frac{1}{2}}.$$

The reader familiar with coordinate geometry will recognize $|p|$ and $|p'|$ as the perpendicular distances of the respective planes from the origin, and that $k = ll' + mm' + nn' = \cos\theta$, where θ is the angle between the two planes whose normals have direction cosines (l, m, n) and (l', m', n'). □

3.4 Envelopes   Families of curves and surfaces often possess loci known as *envelopes*, which are important especially in the theory of differential equations.

If every member of a one-parameter family of plane curves touches a certain curve C in the plane, then C is called the envelope of the family. For example, the family of straight lines

$$x\cos a + y\sin a = 1, \qquad (3.42)$$

where a is the parameter, are all at unit perpendicular distance from the origin. Therefore, all the straight lines touch the circle $x^2 + y^2 = 1$, which is the envelope of (3.42).

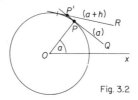

Fig. 3.2

The lines $P'Q$, $P'R$ in Fig. 3.2 correspond to parameter values $a, a+h$. The point of intersection P' is given by the simultaneous equations

$$\phi(x, y, a) \equiv x\cos a + y\sin a - 1 = 0, \qquad (3.43)$$
$$\phi(x, y, a+h) = 0. \qquad (3.44)$$

Subtracting and dividing by h, we get by the **mean-value theorem**,

$$h^{-1}[\phi(x, y, a+h) - \phi(x, y, a)] = \phi_a(x, y, a+\theta h) = 0, \qquad (3.45)$$

where $0 < \theta < 1$. On taking the limit as h tends to zero, we get the limiting position of P', which is the point of contact P of the member a with the envelope C. Thus, the coordinates of P are given by the simultaneous equations

$$\phi(x, y, a) = 0, \qquad \phi_a(x, y, a) = 0. \qquad (3.46)$$

The equation of the envelope is obtained by eliminating a between these two equations. This result applies generally to a family $\phi(x, y, a) = 0$. For example, in the case (3.43), the second of (3.46) is

$$-x\sin a + y\cos a = 0,$$

so that $y = x\tan a$. Substituting for y in (3.43) we find $x = \cos a$, and so the coordinates of P are $x = \cos a$, $y = \sin a$. Eliminating a gives the envelope $x^2 + y^2 = 1$.

Problem 3.15   Determine the envelope of the family of circles
$$x^2 + y^2 - 2ay + \tfrac{1}{2}a^2 = 0, \qquad a > 0.$$
Solution. The equation of the family can be written
$$\phi(x, y, a) = x^2 + (y - a)^2 - \tfrac{1}{2}a^2 = 0, \qquad (3.47)$$
which shows that the member with parameter a has centre $(0, a)$ and radius $a/\sqrt{2}$. Corresponding to (3.46), we have (3.47) and
$$\phi_a = 2(y - a) - a = 0. \qquad (3.48)$$
The latter gives $a = 2y$, and so the elimination of a between (3.47) and (3.48) gives
$$x^2 - y^2 = 0$$
as the equation of the envelope. It is the pair of straight lines $y = \pm x$, which pass through the origin and are inclined at $45°$ to each coordinate axis.   ◻

Problem 3.16   Describe the envelope of the family of spheres
$$(x - a)^2 + [y - f(a)]^2 + z^2 = 1, \qquad (3.49)$$
where f is a given function. Obtain its equation when $f(a) = 2a + 1$.

Solution. The family consists of unit spheres, the centre of the member with parameter a being the point $x = a$, $y = f(a)$, $z = 0$. Thus, the centres all lie on the curve $y = f(x)$ in the xy plane. As a varies, a tube with unit circular cross-section is traced out by the sphere with parameter a. This tubular surface is touched by every member of the family (3.49), and is called the envelope of the family in analogy with envelopes of plane curves.

The equation of the envelope is obtained by eliminating a between
$$\phi(x, y, z, a) = (x - a)^2 + [y - f(a)]^2 + z^2 - 1 = 0, \qquad (3.50)$$
and
$$\phi_a(x, y, z, a) = -2(x - a) - 2[y - f(a)] f'(a) = 0. \qquad (3.51)$$
The elimination is impracticable for general $f(a)$. When $f(a) = 2a + 1$, the last two equations reduce to
$$(x - a)^2 + (y - 2a - 1)^2 + z^2 - 1 = 0, \qquad (3.52)$$
$$x - a + 2(y - 2a - 1) = 0. \qquad (3.53)$$
Solving the latter for a, and substituting in the former, we find for the envelope
$$4x^2 + y^2 + 5z^2 - 4xy + 4x - 2y - 4 = 0,$$
or
$$[(y - 2x - 1)/\sqrt{5}]^2 + z^2 = 1.$$
The last form exhibits the nature of the envelope, since the left-hand side is the square of the distance of the point (x, y, z) from the line

$y-2x-1 = 0$, $z = 0$. The envelope is therefore a unit circular cylinder with this line as axis, as also follows from the above general comments. □

A two-parameter family of surfaces may also possess an envelope. Suppose, for example, that we were to replace $f(a)$ by b in (3.49), where b is a parameter independent of a. Then the new equation would represent a sphere of unit radius centred at $(a, b, 0)$. Every such sphere touches the pair of planes $x = \pm 1$, which is the envelope of the two-parameter family.

In general, the envelope of the family $\phi(x, y, z, a, b) = 0$ is obtained (when it exists) by eliminating a and b between the equations

$$\phi = 0, \quad \phi_a = 0, \quad \phi_b = 0. \tag{3.54}$$

Problem 3.17 Find the envelope of the two-parameter family of planes $z = ax + by + a^2 + b^2$.

Solution. Equations (3.54) are in this case

$$z - ax - by - a^2 - b^2 = 0, \quad -x - 2a = 0, \quad -y - 2b = 0.$$

From the last two we find $a = -\tfrac{1}{2}x$, $b = -\tfrac{1}{2}y$, and substituting in the first we get the required equation

$$z = -\tfrac{1}{4}(x^2 + y^2),$$

which shows that the envelope is the paraboloid of revolution obtained by rotating the parabola $z = -\tfrac{1}{4}x^2$, $y = 0$ about the z axis. □

EXERCISES

1. Expand $xy^3 + x^2 + xy - 1$ in powers of $x-1$ and $y+2$ by means of Taylor's theorem.

2. Expand $(1 - x^2 - y^2)^{-\frac{1}{2}}$ as a Taylor series in powers of x and y as far as terms of fourth degree. Check using the binomial theorem.

3. Find and identify the critical points of $f(x, y) = e^{-x^2}(2xy + y^2)$.

4. Find and identify the critical points of $u = 5x^2 + 2y^2 + z^2 + 2xy + 2z$ using the generalized D test (3.23).

5. Find the least and greatest distances from the origin to the surface $(x/a)^4 + (y/b)^4 + (z/c)^4 = 1$, where $a > b > c > 0$.

6. Given that $u = x^2 + 2y^2 + z^2 + t^2$ possesses a minimum under the constraints $2x - y + z + 2t = 4$, $x + 3y - z + t = 2$, find its location.

7. Find the envelope of the family of surfaces $ax + (y-a)^2 + (z+a)^2 = 0$.

8. Find the envelope of the family of all spheres through the origin with centres on the parabola $y^2 = 4x$, $z = 0$.

9. Obtain the envelope of the two-parameter family of planes $lx + my + nz = 1$, where l, m and n are positive constants satisfying $lmn = \frac{1}{27}$.

Chapter 4

Multiple Integrals

4.1 Double and Repeated Integrals Consider a function $f(x, y)$ which is defined in a closed bounded region R of the xy plane. Let R be divided into n subregions of which the ith has area ΔA_i ($i = 1, 2, ..., n$), and let f_i denote the value of f at an arbitrary point (x_i, y_i) in the ith subregion. If the sum

$$\sum_{i=1}^{n} f_i \, \Delta A_i$$

approaches a finite limit as the dimensions of each subregion become vanishingly small (and n tends to infinity), the limit is called the *double integral* of $f(x, y)$ over R. Here, it is assumed that the limit is independent of the manner in which R is subdivided, and of the particular choice of (x_i, y_i). (This is always the case when f is a *continuous* function.) The double integral is written

$$\iint_R f(x, y) \, dA = \lim_{n \to \infty} \sum_{i=1}^{n} f_i \, \Delta A_i. \tag{4.1}$$

As an example, suppose that a thin metal plate occupies the region R, and that the mass density (mass per unit area) in the vicinity of the point (x, y) is $\sigma(x, y)$. Then for the plate we have:

(i) area $= \iint_R dA$,

(ii) mass $= \iint_R \sigma \, dA$,

(iii) moment of inertia about $Oy = \iint_R x^2 \sigma dA$,

(iv) polar moment of inertia about $O = \iint_R (x^2 + y^2) \sigma \, dA$,

and so on.

If the subdivision of R is rectangular as in Fig. 4.1, the common area of each complete rectangle being $\Delta A_i = \Delta x \, \Delta y$ say, we are led to write the double integral (4.1) as

$$I = \iint_R f(x, y) \, dx \, dy. \tag{4.2}$$

The summation over rectangles may be carried out first over a strip parallel to Oy, followed by a summation over all such strips. This pro-

52

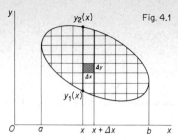

Fig. 4.1

cedure leads to an expression for I as a *repeated integral*. Suppose that each line $x = $ const. which crosses the boundary of R does so in just two points, where $y = y_1(x)$, and $y = y_2(x)$, $(y_1 < y_2)$. Then

$$I = \int_a^b \left\{ \int_{y_1(x)}^{y_2(x)} f(x, y)\, dy \right\} dx, \qquad (4.3)$$

where a and b are the least and greatest values of x in R.

Similarly, if each line $y = $ const. which crosses the boundary of R does so in just two points, where $x = x_1(y)$ and $x = x_2(y)$, $(x_1 < x_2)$, then by interchanging the roles of x and y in the above, we get an alternative form of repeated integral

$$I = \int_c^d \left\{ \int_{x_1(y)}^{x_2(y)} f(x, y)\, dx \right\} dy, \qquad (4.4)$$

where c and d are the least and greatest value of y in R.

If R does not have the properties assumed for (4.3) and (4.4) to apply, it may be possible to divide R into a finite number of parts such that the contribution to I from each part is separately of the form (4.3) or (4.4). This permits the expression of I as a sum of such terms.

Brackets may be omitted in (4.3) and (4.4). Some authors interchange the order of dx and dy on removing the brackets; that convention is not adopted here.

Problem 4.1 If R is the rectangle $0 \leqslant x \leqslant 1$, $-1 \leqslant y \leqslant 2$, express the double integral of $x^2 y^2 (x^2 - y^3)$ over R as a repeated integral in two different ways, and evaluate each.

Solution. If the y integration is carried out first, with x constant, we get

$$I = \iint_R x^2 y^2 (x^2 - y^3)\, dx\, dy = \int_0^1 \left\{ \int_{-1}^2 x^2 y^2 (x^2 - y^3)\, dy \right\} dx$$

$$= \int_0^1 \left\{ \int_{-1}^2 (x^4 y^2 - x^2 y^5)\, dy \right\} dx$$

$$= \int_0^1 \left| \tfrac{1}{3} x^4 y^3 - \tfrac{1}{6} x^2 y^6 \right|_{y=-1}^{y=2} dx$$

$$= \int_0^1 (3x^4 - \tfrac{21}{2} x^2)\, dx = -\tfrac{29}{10}.$$

53

If the x integration is carried out first, with y constant,

$$I = \int_{-1}^{2} \left\{ \int_{0}^{1} (x^4 y^2 - x^2 y^5)\, dx \right\} dy$$
$$= \int_{-1}^{2} \left| \tfrac{1}{5}x^5 y^2 - \tfrac{1}{3}x^3 y^5 \right|_{x=0}^{x=1} dy$$
$$= \int_{-1}^{2} (\tfrac{1}{5}y^2 - \tfrac{1}{3}y^5)\, dy = -\tfrac{29}{10},$$

as before. □

Problem 4.2 Find the mass of a thin metal plate lying in the positive quadrant of the xy plane and bounded by the lines $y = 2x+1$, $y = x^2+1$, if its density is $x^2 y$ at the point (x, y).

Solution. The lines $y = 2x+1$ and $y = x^2+1$ meet at the points $(0, 1)$ and $(2, 5)$. (The reader should make a sketch.) Therefore, the total mass is given by the integral

$$M = \int_{0}^{2} \int_{x^2+1}^{2x+1} x^2 y\, dy\, dx = \int_{0}^{2} \left| \tfrac{1}{2}x^2 y^2 \right|_{y=x^2+1}^{y=2x+1} dx$$
$$= \tfrac{1}{2} \int_{0}^{2} [x^2(2x+1)^2 - x^2(x^2+1)^2]\, dx = \tfrac{184}{35}. \quad □$$

Problem 4.3 Evaluate

$$I = \int_{0}^{a} \int_{x}^{a} \frac{x}{\sqrt{(x^2+y^2)}}\, dy\, dx \tag{4.5}$$

by first changing the order of integration.

Solution. We note that the integral is difficult to evaluate without changing the order of integration, for since

$$\frac{\partial}{\partial y} \sinh^{-1}\left(\frac{y}{x}\right) = \frac{1}{\sqrt{(x^2+y^2)}},$$

we should have

$$I = \int_{0}^{1} \left| x \sinh^{-1}(y/x) \right|_{y=x}^{y=a} dx$$
$$= \int_{0}^{a} [x \sinh^{-1}(a/x) - x \sinh^{-1} 1]\, dx,$$

which does not permit easy integration.

When the order is changed, the limits need to be changed. The new limits can be found by examining the region R over which the double integral corresponding to I applies. From (4.5), we have that $x \leqslant y \leqslant a$, and so R is bounded below by part of the line $y = x$, and above by part of the line $y = a$. These lines meet at (a, a), and since the limits for x are $0, a$, the region R is as shown in Fig. 4.2.

Any line $y = \text{const.}$ which meets the boundary of R twice, does so where $x = 0$ and $x = y$ (P, Q in the Figure), and so these are the limits

54

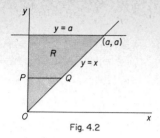

Fig. 4.2

for x when we integrate with respect to this variable first. Thus,

$$I = \int_0^a \int_0^y [x/\sqrt{(x^2+y^2)}] \, dx \, dy$$

$$= \int_0^a \sqrt{(x^2+y^2)} \Big|_{x=0}^{x=y} \, dy$$

$$= \int_0^a (\sqrt{2}y - y) \, dy = \tfrac{1}{2}(\sqrt{2}-1) a^2. \qquad \square$$

Problem 4.4 Change the order of integration in the repeated integral

$$I = \int_0^{2c} \int_{\sqrt{(2cy-y^2)}}^{\sqrt{(2cy)}} \sin xy \, dx \, dy, \qquad (c > 0). \qquad (4.6)$$

Solution. The lower limit for x satisfies the equation

$$x^2 + y^2 - 2cy = 0,$$

i.e.
$$x^2 + (y-c)^2 = c^2, \qquad (4.7)$$

while the upper limit satisfies

$$x^2 = 2cy. \qquad (4.8)$$

Therefore, the region of integration R for the double integral corresponding to I is bounded to the left by part of the circle (4.7), and is bounded to the right by part of the parabola (4.8). Since $0 \leqslant x \leqslant 2c$, R is the shaded region in Fig. 4.3, Q being the point $(2c, 2c)$.

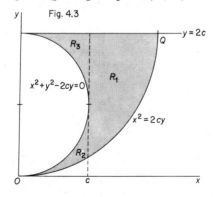

Fig. 4.3

Not all lines parallel to Oy meet R in at most two points, and so we divide R into three parts R_1, R_2', R_3, as shown in the Figure, by means of a cut along the line $x = c$. Let I_1, I_2, I_3, respectively, denote the contributions to I from each of these parts. In R_1, the limits for y are $x^2/2c$, $2c$, while the limits for x are c, $2c$, and so

$$I_1 = \int_c^{2c} \int_{x^2/2c}^{2c} \sin xy \, dy \, dx.$$

By treating the other two contributions to I in a similar way we find, since $I = I_1 + I_2 + I_3$,

$$I = \int_c^{2c} \int_{x^2/2c}^{2c} \sin xy \, dy \, dx + \int_0^c \int_{x^2/2c}^{c - \sqrt{(c^2 - x^2)}} \sin xy \, dy \, dx$$
$$+ \int_0^c \int_{c + \sqrt{(c^2 - x^2)}}^{2c} \sin xy \, dy \, dx. \qquad \square$$

4.2 Transformations of Double Integrals

By a suitable change of variables (transformation)

$$x = x(u, v), \quad y = y(u, v), \tag{4.9}$$

we can express an integral of the form

$$I = \iint_R f(x, y) \, dx \, dy \tag{4.10}$$

as a double integral over the image region S in the uv plane.

Suppose that the functions in (4.9) are continuous and have continuous first derivatives, and that the Jacobian

$$J = \partial(x, y)/\partial(u, v) \neq 0 \tag{4.11}$$

for (u, v) in S. Then (4.9) can be inverted, which shows that a point in the region R can be identified equally well by giving the values of (u, v) as by giving its coordinates (x, y). We call (u, v) *curvilinear coordinates* in the region, because the two families of lines

$$u = \text{const.}, \quad v = \text{const.}, \tag{4.12}$$

are in general curves.

Let R be divided into area elements by a grid of lines $u = \text{const.}$ with common spacing Δu, and lines $v = \text{const.}$ with common spacing Δv. A typical element is approximately a parallelogram whose vertices A, B, C, D have curvilinear coordinates (u, v), $(u + \Delta u, v)$, $(u, v + \Delta v)$, $(u + \Delta u, v + \Delta v)$, respectively, and whose area is

$$AB \, . \, AC \sin BAC = |\mathbf{AB} \wedge \mathbf{AC}| \tag{4.13}$$

in vector notation. Since v is constant along AB, the increments in x and y between these two vertices are approximately

$$x_B - x_A = \frac{\partial x}{\partial u} \Delta u, \qquad y_B - y_A = \frac{\partial y}{\partial u} \Delta u,$$

giving for the rectangular components of **AB**:

$$\mathbf{AB} = \left(\frac{\partial x}{\partial u} \Delta u, \; \frac{\partial y}{\partial u} \Delta u \right).$$

A corresponding result for **AC** is

$$\mathbf{AC} = \left(\frac{\partial x}{\partial v} \Delta v, \; \frac{\partial y}{\partial v} \Delta v \right),$$

and so on forming the vector product in (4.13), we obtain for the area of the elementary parallelogram,

$$\left| \frac{\partial x}{\partial u} \frac{\partial y}{\partial v} - \frac{\partial y}{\partial u} \frac{\partial x}{\partial v} \right| \Delta u \, \Delta v = \left| \frac{\partial(x, y)}{\partial(u, v)} \right| \Delta u \, \Delta v, \tag{4.14}$$

where terms of higher degree than the second in Δu and Δv are neglected.

By carrying out the limiting process in (4.1) with the present subdivision of R, we obtain an equivalent double integral to (4.10),

$$I = \iint\limits_{R} f(x, y) \, dx \, dy = \iint\limits_{S} f\left[x(u, v), y(u, v)\right] \left| \frac{\partial(x, y)}{\partial(u, v)} \right| du \, dv, \tag{4.15}$$

where S is the image, in the uv plane, of the region R. Note that it is the absolute value of the Jacobian (4.11) that appears in (4.15).

In the particular case of the transformation to plane polar coordinates (r, θ), given by $x = r \cos \theta$, $y = r \sin \theta$, we find

$$J = \partial(x, y)/\partial(r, \theta) = r.$$

Hence

$$I = \iint\limits_{S} f(r \cos \theta, r \sin \theta) r \, dr \, d\theta, \tag{4.16}$$

where S is the appropriate region of the $r\theta$ plane. Although the Jacobian vanishes at the single point $x = 0$, $y = 0$, no difficulty arises when R includes this point. We can always regard I as the limit of an integral over a modified region R_1, formed by deleting from R a neighbourhood of the origin. Normally, this again leads to (4.16) without change.

Problem 4.5 Evaluate

$$I = \int_{-a}^{a} \int_{-\sqrt{(a^2 - y^2)}}^{\sqrt{(a^2 - y^2)}} (x^2 + y^2)^{\frac{3}{2}} \, dx \, dy,$$

by changing to polar coordinates.

Solution. Since $-\sqrt{(a^2 - y^2)} \leqslant x \leqslant \sqrt{(a^2 - y^2)}$ is equivalent to $x^2 \leqslant a^2 - y^2$, the region of integration in the xy plane is bounded by the circle of radius a and centre O. The corresponding region of the $r\theta$ plane

57

E

is therefore rectangular, being given by $0 \leqslant r \leqslant a, 0 \leqslant \theta \leqslant 2\pi$. Therefore, by (4.16)

$$I = \int_0^{2\pi} \int_0^a r^3 . r \, dr \, d\theta = \int_0^{2\pi} \int_0^a r^4 \, dr \, d\theta$$

$$= \int_0^{2\pi} \left| \tfrac{1}{5} r^5 \right|_0^a \, d\theta = \tfrac{1}{5} a^5 \int_0^{2\pi} d\theta = 2\pi a^5 / 5. \qquad \square$$

Problem 4.6 Find the area of the region in the positive quadrant of the xy plane bounded by the curves

$$x^2 + 2y^2 = 1, \quad x^2 + 2y^2 = 4, \quad y = 2x, \quad y = 5x.$$

Solution. Put

$$u = x^2 + 2y^2, \qquad v = y/x. \qquad (4.17)$$

Then the image in the uv plane of the given region R is the rectangular region S: $1 \leqslant u \leqslant 4, \; 2 \leqslant v \leqslant 5$. By (4.17),

$$\frac{\partial(u, v)}{\partial(x, y)} = \begin{vmatrix} u_x & v_x \\ u_y & v_y \end{vmatrix} = \begin{vmatrix} 2x & 4y \\ -y/x^2 & 1/x \end{vmatrix} = 2 + 4y^2/x^2,$$

and so

$$\frac{\partial(x, y)}{\partial(u, v)} = \frac{1}{2 + 4y^2/x^2} = \frac{1}{2(1 + 2v^2)}.$$

Thus for R:

$$\text{Area} = \iint_R 1 \, dx \, dy = \iint_S \left| \partial(x, y)/\partial(u, v) \right| \, du \, dv$$

$$= \int_2^5 \int_1^4 \frac{du \, dv}{2(1 + 2v^2)} = \int_2^5 \left| \frac{1}{2\sqrt{2}} \tan^{-1} \sqrt{2} v \right|_1^4 \, du$$

$$= \frac{3}{2\sqrt{2}} (\tan^{-1} 4\sqrt{2} - \tan^{-1} \sqrt{2}) = \frac{3}{2\sqrt{2}} \tan^{-1} \left(\frac{\sqrt{2}}{3} \right). \qquad \square$$

Problem 4.7 Evaluate the integral

$$I = \iint_R \sqrt{[1 - (x^2/a^2) - (y^2/b^2)]} \, dx \, dy, \qquad a > 0, b > 0, \qquad (4.18)$$

where R is the region enclosed by the ellipse $(x^2/a^2) + (y^2/b^2) = 1$.

Solution. The form of integrand, and the region itself, suggest the substitution

$$x/a = u \cos v, \qquad y/b = u \sin v. \qquad (4.19)$$

The transformed region is the rectangle $0 \leqslant u \leqslant 1, \; 0 \leqslant v \leqslant 2\pi$. The Jacobian of the transformation is easily found to be

$$\partial(x, v)/\partial(u, v) = abu,$$

58

and hence by (4.19) we get

$$I = \int_0^{2\pi} \int_0^1 \sqrt{(1 - u^2)} \,.\, abu \, du \, dv$$

$$= ab \int_0^{2\pi} \left| -\tfrac{1}{3}(1 - u^2)^{\frac{3}{2}} \right|_0^1 dv$$

$$= \tfrac{1}{3}ab \int_0^{2\pi} 1 \,.\, dv = \tfrac{2}{3}\pi ab.$$

□

Problem 4.8   Evaluate

$$I = \int_0^\infty e^{-x^2} dx, \tag{4.20}$$

by expressing I^2 in the form of a double integral and transforming to polar coordinates.

Solution. The integral is *improper* in that the range of integration is infinite. In manipulation, we must keep in mind that the notation in (4.20) has the interpretation

$$I = \lim_{a \to \infty} I_a = \lim_{a \to \infty} \int_0^a e^{-x^2} dx, \tag{4.21}$$

assuming that the limit exists. Since x acts only as a 'dummy' variable in (4.21), we can equally well write

$$I_a = \int_0^a e^{-y^2} dy,$$

and so, because I_a depends only on a, we have

$$I_a^2 = \int_0^a I_a e^{-y^2} dy = \int_0^a \left(\int_0^a e^{-x^2} dx \right) e^{-y^2} dy$$

$$= \int_0^a \left(\int_0^a e^{-x^2} e^{-y^2} dx \right) dy = \int_0^a \int_0^a e^{-(x^2 + y^2)} dx \, dy.$$

Therefore, if R_a denotes the square region $0 \leqslant x \leqslant a, 0 \leqslant y \leqslant a$, we have an expression for I^2 in terms of a (limiting) double integral.

$$I^2 = \lim_{a \to \infty} I_a^2 = \lim_{a \to \infty} \iint_{R_a} e^{-(x^2 + y^2)} dx \, dy.$$

The transformation to polar coordinates (r, θ), where $x = r \cos \theta$, $y = r \sin \theta$, leads to the replacement of $dx \, dy$ by $r \, dr \, d\theta$ (cf. (4.16), and therefore

$$I_a^2 = \iint_{S_a} e^{-r^2} r \, dr \, d\theta, \tag{4.22}$$

where S_a is the image, in the $r\theta$ plane, of R_a. The limits here will be complicated, because a square region is not well suited for the introduction of polar coordinates.

But note that R_a contains the quarter-circular region C_a: $x^2 + y^2 \leqslant a^2$,

$x \geqslant 0$, $y \geqslant 0$, and since $e^{-(x^2+y^2)}$ is positive, its double integral over C_a is less than I_a^2. Likewise, I_a^2 is less than the double integral of $e^{-(x^2+y^2)}$ over $C_{\sqrt{2}a}$ (which contains R_a). However, as a tends to infinity, the double integrals over C_a and $C_{\sqrt{2}a}$ must tend to the same limit, and so this limit must be I^2. Therefore,

$$I^2 = \lim_{a \to \infty} \int_0^{\frac{1}{2}\pi} \int_0^a e^{-r^2} r \, dr \, d\theta$$

$$= \lim_{a \to \infty} \int_0^{\frac{1}{2}\pi} \left| -\tfrac{1}{2} e^{-r^2} \right|_0^a \, d\theta$$

$$= \lim_{a \to \infty} \tfrac{1}{4}\pi(1 - e^{-a^2}) = \tfrac{1}{4}\pi,$$

whence $I = \tfrac{1}{2}\sqrt{\pi}$.

This is one of several methods of evaluating the important integral (4.20), and succeeds through the introduction of the factor r as the Jacobian of the transformation to polar coordinates. $\quad\square$

4.3 Triple Integrals Let $f(x, y, z)$ be defined in a closed bounded three-dimensional region R, and let R be divided into n subregions of which the ith has volume ΔV_i ($i = 1, 2, \ldots, n$). If f_i denotes the value of f at an arbitrary point in the ith subregion, then the *triple integral* of f over R is

$$I = \iiint_R f(x, y, z) \, dV = \lim_{n \to \infty} \sum_{i=1}^{n} f_i \Delta V_i, \qquad (4.23)$$

provided that a definite limit exists on the right as the dimensions of all ΔV_i tend to zero (independently of the manner of subdivision). This is always the case when f is continuous.

The triple integral can be expressed as a repeated integral just as a double integral can.

Problem 4.9 Find (i) the total mass, (ii) the moment of inertia about Ox, of a body which occupies the positive octant of the sphere $x^2 + y^2 + z^2 \leqslant 1$ and has mass density $\rho(x, y, z) = x^3 yz$.

Solution. (i) The mass density ρ is the mass per unit volume of the body in the vicinity of the point (x, y, z), and so if R denotes the region occupied by the body, its total mass is

$$M = \iiint_R \rho \, dV. \qquad (4.24)$$

As indicated in Fig. 4.4, let the summation over such blocks be carried out first for those which form a rectangular prism parallel to Ox (y and z being constant). On the prism we have the inequality $0 \leqslant x \leqslant \sqrt{(1 - y^2 - z^2)}$. Next, sum over all such prisms which form a slice parallel

60

to the xy plane (z being constant). On the slice, we have $0 \leqslant y \leqslant \sqrt{(1-z^2)}$. Finally, sum over all slices, $0 \leqslant z \leqslant 1$. This procedure leads to the expression of (4.24) as the repeated integral

$$
\begin{aligned}
M &= \int_0^1 \int_0^{\sqrt{(1-z^2)}} \int_0^{\sqrt{(1-y^2-z^2)}} x^3 yz \, dx \, dy \, dz \\
&= \int_0^1 \int_0^{\sqrt{(1-z^2)}} \tfrac{1}{4}x^4 \Big|_0^{\sqrt{(1-y^2-z^2)}} yz \, dy \, dz \\
&= \tfrac{1}{4} \int_0^1 \int_0^{\sqrt{(1-z^2)}} (1-y^2-z^2)^2 yz \, dy \, dz \\
&= \tfrac{1}{4} \int_0^1 \left| -\tfrac{1}{6}(1-y^2-z^2)^3 \right|_{y=0}^{\sqrt{(1-z^2)}} z \, dz \\
&= \tfrac{1}{24} \int_0^1 (1-z^2)^3 z \, dz \\
&= \tfrac{1}{24} \left| -\tfrac{1}{8}(1-z^2)^4 \right|_0^1 = \tfrac{1}{192}.
\end{aligned}
$$

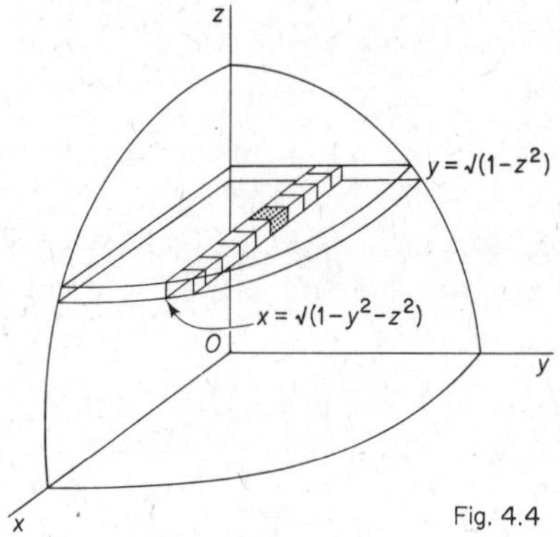

$y = \sqrt{(1-z^2)}$

$x = \sqrt{(1-y^2-z^2)}$

Fig. 4.4

(ii) The moment of inertia about Ox is

$$
\begin{aligned}
I_x &= \iiint_R (y^2+z^2)\rho \, dV \\
&= \int_0^1 \int_0^{\sqrt{(1-z^2)}} \int_0^{\sqrt{(1-y^2-z^2)}} (y^2+z^2)x^3 yz \, dx \, dy \, dz \\
&= \int_0^1 \int_0^{\sqrt{(1-z^2)}} \tfrac{1}{4}x^4 \Big|_0^{\sqrt{(1-y^2-z^2)}} (y^2+z^2) \, yz \, dy \, dz \\
&= \tfrac{1}{4} \int_0^1 \int_0^{\sqrt{(1-z^2)}} (1-y^2-z^2)^2 (y^2+z^2) \, yz \, dy \, dz \\
&= \tfrac{1}{4} \int_0^1 \int_0^{\sqrt{(1-z^2)}} [(1-y^2-z^2)^2 - (1-y^2-z^2)^3] \, yz \, dy \, dz.
\end{aligned}
$$

This repeated integral can be expressed as the difference of two integrals, corresponding to each of the terms appearing in the square bracket. The

61

first integral has already been evaluated in (i). The second is evaluated in like manner, and we find

$$\tfrac{1}{4} \int_0^1 \int_0^{\sqrt{(1-z^2)}} (1-y^2-z^2)^3 yz \, dy \, dz = \tfrac{1}{320},$$

and so

$$I_x = \tfrac{1}{192} - \tfrac{1}{320} = \tfrac{2}{5}. \qquad \blacksquare$$

4.4 Transformations of Triple Integrals   If

$$I = \iiint\limits_{R} f(x, y, z) \, dx \, dy \, dz, \qquad (4.25)$$

and we change variables in accordance with the transformation

$$x = x(u, v, w), \quad y = y(u, v, w), \quad z = z(u, v, w), \qquad (4.26)$$

(4.25) becomes

$$I = \iiint\limits_{S} F(u, v, w) \left| \frac{\partial(x, y, z)}{\partial(u, v, w)} \right| du \, dv \, dw, \qquad (4.27)$$

where S is the image in uvw space of the region R, and

$$F(u, v, w) = f[x(u, v, w), y(u, v, w), z(u, v, w)].$$

It is assumed here that the functions (4.26) are continuous with continuous partial derivatives, and that the Jacobian $\partial(x, y, z)/\partial(u, v, w)$ is of one sign in the region of integration.

The transformations to three-dimensional polar coordinates are important.

Cylindrical polar coordinates (Fig. 4.5):

$$x = \rho \cos \phi, \quad y = \rho \sin \phi, \quad z = z, \qquad (4.28)$$

$$J = \partial(x, y, z)/\partial(\rho, \phi, z) = \rho.$$

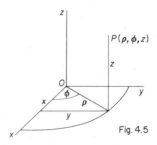

Fig. 4.5

Spherical polar coordinates (Fig. 4.6):

$$x = r \sin \theta \cos \phi, \quad y = r \sin \theta \sin \phi, \quad z = r \cos \theta, \qquad (4.29)$$

$$J = \partial(x, y, z)/\partial(r, \theta, \phi) = r^2 \sin \theta.$$

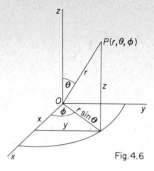

Fig. 4.6

Problem 4.10 Evaluate

$$I = \iiint\limits_{R} xyz(x^2 + y^2 + z^2)^{-\frac{1}{2}} \, dx \, dy \, dz.$$

where R is the positive octant of the sphere $x^2 + y^2 + z^2 \leqslant a^2$.

Solution. Transforming to spherical polar coordinates, we get

$$I = \int_0^{\frac{1}{2}\pi} \int_0^{\frac{1}{2}\pi} \int_0^a r^2 \sin^2\theta \cos\theta \cos\phi \sin\phi \left| r^2 \sin\theta \right| dr \, d\theta \, d\phi$$

$$= \int_0^{\frac{1}{2}\pi} \int_0^{\frac{1}{2}\pi} \left| \tfrac{1}{5}r^5 \right|_0^a \sin^3\theta \cos\theta \cos\phi \sin\phi \, d\theta \, d\phi$$

$$= \tfrac{1}{5}a^5 \int_0^{\frac{1}{2}\pi} \left| \tfrac{1}{4}\sin^4\theta \right|_0^{\frac{1}{2}\pi} \cos\phi \sin\phi \, d\phi$$

$$= \tfrac{1}{20}a^5 \left| \tfrac{1}{2}\sin^2\phi \right|_0^{\frac{1}{2}\pi} = a^5/40. \qquad \square$$

Problem 4.11 Find the volume of the uppermost region enclosed by the surface of revolution $4(x^2 + y^2) = z^4$ and the sphere $x^2 + y^2 + z^2 = 3$.

Solution. In terms of cylindrical polar coordinates, the equation $4(x^2 + y^2) = z^4$ becomes $2\rho = z^2$, which shows that it represents the surface obtained by rotating the parabola $y = \tfrac{1}{2}z^2$, $x = 0$ (whose axis is Oy, and vertex the origin), about Oz.

The equation of the sphere in cylindrical polars is $\rho^2 = 3 - z^2$, and the two surfaces intersect where $3 - \rho^2 = 2\rho$. Since $\rho \geqslant 0$, the equations of the two horizontal circles of intersection are therefore

$$\rho = 1, \qquad z = \pm\sqrt{2}.$$

In the region of interest R, we have

$$\sqrt{(2\rho)} \leqslant z \leqslant \sqrt{(3 - \rho^2)},$$

63

and the volume of R is

$$\iiint_R dx\,dy\,dz = \int_0^1 \int_0^{2\pi} \int_{\sqrt{(2\rho)}}^{\sqrt{(3-\rho^2)}} \rho\,dz\,d\phi\,d\rho$$

$$= \int_0^1 \int_0^{2\pi} \rho \left|z\right|_{\sqrt{(2\rho)}}^{\sqrt{(3-\rho^2)}} d\phi\,d\rho$$

$$= \int_0^1 \int_0^{2\pi} [\rho(3-\rho^2)^{\frac{1}{2}} - \sqrt{2}\rho^{\frac{3}{2}}]\,d\phi\,d\rho$$

$$= 2\pi \left| -\frac{1}{3}(3-\rho^2)^{\frac{3}{2}} - \frac{2\sqrt{2}}{5}\rho^{\frac{5}{2}} \right|_0^1$$

$$= 2\pi\left(-\frac{2\sqrt{2}}{3} + \sqrt{3} - \frac{2\sqrt{2}}{5}\right) = \frac{2\pi}{15}(15\sqrt{3} - 16\sqrt{2}). \qquad \square$$

Problem 4.12 Evaluate the integral

$$\iiint_R [1-(x/a)^2 -(y/b)^2 -(z/c)^2]^{\frac{3}{2}}\,dx\,dy\,dz, \qquad (4.30)$$

where R is the region enclosed by the ellipsoid $(x/a)^2 + (y/b)^2 + (z/c)^2 = 1$.
 Solution. Introduce a modified type of polar coordinates by putting

$$x/a = r\sin\theta\cos\phi, \quad y/b = r\sin\theta\sin\phi, \quad z/c = r\cos\theta. \qquad (4.31)$$

The region R is given in these coordinates by

$$0 \leqslant r \leqslant 1, \quad 0 \leqslant \theta \leqslant \pi, \quad 0 \leqslant \phi < 2\pi,$$

and the Jacobian of the transformation (4.31) is found by direct calculation to be

$$\partial(x, y, z)/\partial(r, \theta, \phi) = abcr^2\sin\theta.$$

Therefore, the integral (4.30) transforms to

$$\int_0^1 \int_0^\pi \int_0^{2\pi} (1-r^2)^{\frac{3}{2}} abcr^2\sin\theta\,d\phi\,d\theta\,dr = 2\pi abc \int_0^1 \int_0^\pi (1-r^2)^{\frac{3}{2}} r^2\sin\theta\,d\theta\,dr$$

$$= 2\pi abc \int_0^1 (1-r^2)^{\frac{3}{2}} r^2 \left| -\cos\theta \right|_0^\pi dr$$

$$= 4\pi abc \int_0^1 (1-r^2)^{\frac{3}{2}} r^2\,dr.$$

The substitution $r = \sin\alpha$ reduces the last integral to a standard elementary form, giving for the integral (4.30)

$$4\pi abc \int_0^{\frac{1}{2}\pi} \cos^4\alpha \sin^2\alpha\,d\alpha = \tfrac{1}{8}\pi^2 abc. \qquad \square$$

Problem 4.13 Evaluate the integral

$$I = \iiint_R (x+y+z)(x+y-z)(x-y-z)\,dx\,dy\,dz,$$

where R is the tetrahedron bounded by the planes $x+y+z = 0$, $x+y-z = 0$, $x-y-z = 0$, $2x-z = 1$.

64

Solution. The form of the integral and the equations of three of the bounding planes of R suggest the transformation

$$u = x+y+z, \quad v = x+y-z, \quad w = x-y-z, \qquad (4.32)$$

for which the Jacobian is

$$\frac{\partial(u, v, w)}{\partial(x, y, z)} = \begin{vmatrix} 1 & 1 & 1 \\ 1 & 1 & -1 \\ 1 & -1 & -1 \end{vmatrix} = -4,$$

so that $|\partial(x, y, z)/\partial(u, v, w)| = \frac{1}{4}$. The equation $2x - z = 1$ becomes

$$(u+w) - \tfrac{1}{2}(u-v) = 1,$$

i.e.
$$u + v + 2w = 2,$$

and so the image region S in the uvw plane, of the region R under the transformation (4.32), is the tetrahedron bounded by the planes

$$u = 0, \quad v = 0, \quad w = 0, \quad u+v+2w = 2.$$

The last of these planes meets the u, v, and w axes at the points $(2, 0, 0)$, $(0, 2, 0)$, $(0, 0, 1)$, respectively, and therefore S is the tetrahedron in Fig. 4.7.

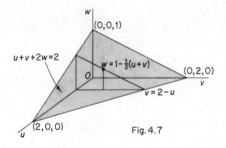

Fig. 4.7

To find the new limits of integration, consider an arbitrary plane $u = \text{const.}$ which cuts S. Along any line $v = \text{const.}$, intercepted by S, we have $0 \leqslant w \leqslant 1 - \tfrac{1}{2}(u+v)$. Here, v must satisfy $0 \leqslant v \leqslant 2-u$, while $0 \leqslant u \leqslant 2$. Therefore,

$$I = \int_0^2 \int_0^{2-u} \int_0^{1-\frac{1}{2}(u+v)} \tfrac{1}{4} uvw \, dw \, dv \, du.$$

The integration is now straightforward, and is found to lead to the value

$$I = 1/180.$$

This value may, of course, have been obtained by performing the integration in any other order, with the introduction of appropriate limits. ☐

65

EXERCISES

1. Evaluate the integrals

   (i) $\int_0^2 \int_0^3 (2x + y)^2 \, dy \, dx.$ (ii) $\int_0^{\frac{1}{4}\pi} \int_{\cos\theta}^1 r \cos\theta \, dr \, d\theta.$

   Sketch the regions in the xy plane over which the corresponding double integrals are taken, if $x = r\cos\theta$, $y = r\sin\theta$ in (ii).

2. Evaluate $\int_0^1 \int_x^{\sqrt{x}} (x^2 + y^2) \, dy \, dx$. Sketch the region for the corresponding double integral, and check your evaluation by reversing the order of integration.

3. Use the transformation $u = x - 2y$, $v = x + 2y$ to evaluate

   $$\int_0^1 \int_0^{2-2y} \exp\left[(x - 2y)/(x + 2y)\right] dx \, dy.$$

4. Evaluate

   $$\iint_R x^3 y^3 \sqrt{(1 - x^4 - y^4)} \, dx \, dy.$$

   where R is the region $x \geqslant 0$, $y \geqslant 0$, $x^4 + y^4 \leqslant 1$, by using the transformation $x^2 = u\cos v$, $y^2 = u\sin v$.

5. Find the volume of the finite region lying between the surfaces $z = 1 - x^2 - y^2$ and $z = 13 - 4(x^2 + y^2)$.

6. Show that the centroid of the positive octant ($x \geqslant 0$, $y \geqslant 0$, $z \geqslant 0$) of the solid sphere $x^2 + y^2 + z^2 \leqslant a^2$ is at the point $\bar{x} = \bar{y} = \bar{z} = \frac{3}{8}a$. (The centroid $(\bar{x}, \bar{y}, \bar{z})$ of a three-dimensional region R is defined by

   $$V\bar{x} = \iiint x \, dx \, dy \, dz,$$

   with corresponding forms for $\bar{y}, \bar{z}$, where V is the volume and the integration is taken over R.)

7. Evaluate

   $$\iiint_R x(x + y + z) \, dx \, dy \, dz,$$

   where R is the region bounded by the planes $x = 0$, $z = 0$, $y - z = 0$, $x + y + z = 1$.

Chapter 5

Line and Surface Integrals

5.1 Line Integrals   In this chapter we shall mean by the word *curve* a one-parameter set of points $P(x, y, z)$ defined by relations of the form

$$x = x(t), \quad y = y(t), \quad z = z(t), \tag{5.1}$$

where t takes all values in some interval $a \leqslant t \leqslant b$, and the functions appearing in (5.1) are continuous. A curve which meets or crosses itself does so at *multiple points*, i.e. points (x, y, z) corresponding to more than one value of t. If the points corresponding to $t = a$ and $t = b$ coincide, the curve is *closed*.

A curve with no multiple points is said to be *simple*; if a curve is closed and has no multiple points other than the coincident end-points, it is a *simple closed curve*.

When the functions in (5.1) have derivatives, and these do not all vanish at the same point, the curve possesses a tangential direction which has the direction of the vector with these derivatives as components:

$$[x'(t), y'(t), z'(t)]. \tag{5.2}$$

If the tangential direction varies continuously from point to point, i.e. if the components (5.1) are continuous, the curve is said to be *smooth*.

Problem 5.1   Find the length of the curve C:

$$x = 3 \cos t, \quad y = 3 \sin t, \quad z = t^2, \quad (0 \leqslant t \leqslant 1).$$

Solution. Consider first a general smooth curve of the form (5.1). Let it be approximated by a polygonal line of n segments joining points with parameters $t_0 = a, t_1, t_2, \ldots, t_{n-1}, t_n = b$. The segment (t_k, t_{k+1}) has length

$$\Delta s_k = [(\Delta x_k)^2 + (\Delta y_k)^2 + (\Delta z_k)^2]^{\frac{1}{2}}, \tag{5.3}$$

where $\Delta x_k = x(t_{k+1}) - x(t_k)$, and Δy_k and Δz_k are correspondingly defined. By the mean-value theorem, we have

$$\Delta x_k = x'(\tau_k)(t_{k+1} - t_k),$$

where τ_k lies between t_k and t_{k+1}, together with two similar expressions for the y and z increments. Substituting in (5.3), summing over all segments and taking a limiting process as $t_{k+1} - t_k \to 0$ and $n \to \infty$, leads to the integral

$$\int_a^b [x'^2(t) + y'^2(t) + z'^2(t)]^{\frac{1}{2}} \, dt, \tag{5.4}$$

which is defined to be the length of the smooth curve (5.1).

For the curve C in question, we have

$$x'(t) = -3\sin t, \quad y'(t) = 3\cos t, \quad z'(t) = 2t, \tag{5.5}$$

whence (5.4) gives for the required length

$$\int_0^1 (9+4t^2)^{\frac{1}{2}} \, dt = \frac{1}{4}\left[2\sqrt{13}+9\ln\left(\frac{2+\sqrt{13}}{3}\right)\right]. \qquad \square$$

If s denotes arc length along the curve measured from the end-point $t = a$, then

$$ds/dt = [x'^2(t)+y'^2(t)+z'^2(t)]^{\frac{1}{2}}, \tag{5.6}$$

and (5.4) can be written

$$\int_C ds.$$

If $f(x, y, z)$ is defined at all points on C, and each polygonal segment is multiplied by the value of f at an arbitrary point on the arc intercepted by the segment, then the limiting process leading to (5.4) gives

$$\int_C f \, ds = \int_a^b f[x(t), y(t), z(t)] \, \frac{ds}{dt} \, dt. \tag{5.7}$$

Problem 5.2 Evaluate the line integral

$$\int_C (y+xy^{-1}+2yz) \, ds,$$

where C is the curve $x = t^2$, $y = t$, $z = 1$, $0 \leqslant t \leqslant 1$.

Solution. The term *line integral* is used to emphasize the fact that the integration is performed along a specified curve or line. By (5.6), and the equations defining C,

$$ds/dt = [(2t)^2+(1)+(0)]^{\frac{1}{2}} = (4t^2+1)^{\frac{1}{2}}.$$

Therefore, substituting for x, y and z in the integrand, in terms of t, we get for the required integral

$$\int_0^1 (t+t+2t)(4t^2+1)^{\frac{1}{2}} \, dt = 4\int_0^1 t(4t^2+1)^{\frac{1}{2}} \, dt = (5\sqrt{5}-1)/3. \qquad \square$$

The form (5.7) is not the only form of line integral of a function $f(x, y, z)$ over a curve C. Let C be approximated by a polygonal line of n segments, the x coordinates of the end-points of a typical segment being x_k, x_{k+1}, and let f_k be the value of f at an arbitrary point of the arc of C intercepted by this segment. If the sum over all such segments

$$\sum_k f_k(x_{k+1}-x_k)$$

tends to a definite limit as $x_{k+1} - x_k \to 0$ and $n \to \infty$, the limit is denoted by

$$\int_C f(x, y, z)\, dx.$$

Similarly, line integrals with respect to y and z are defined.

Problem 5.3   Evaluate

$$\int_C (x^2 y^2 + xy - yz)\, dx, \tag{5.8}$$

where C is the semicircular arc of the circle $x^2 + y^2 = 1$, $z = 0$, for which $y \geqslant 0$ and x increases from -1 to 1.

Solution. The third term in the integrand vanishes on C since $z = 0$. In the other two terms we substitute $y = \sqrt{(1-x^2)}$, and hence obtain for (5.8)

$$\int_{-1}^{1} [x^2(1-x^2) + x\sqrt{(1-x^2)}]\, dx = \left| \tfrac{1}{3}x^3 - \tfrac{1}{5}x^5 - \tfrac{1}{3}(1-x^2)^{\frac{3}{2}} \right|_{-1}^{1}$$

$$= \tfrac{2}{3} - \tfrac{2}{5} = \tfrac{4}{15}.$$

Note that if the integration were performed along the same curve in the opposite sense, the limits in the last integral would be interchanged, giving $-\tfrac{4}{15}$ for the value.

Alternatively, express the equation of C in the parametric form $x = \cos t$, $y = \sin t$, $z = 0$, where t increases from $-\pi$ to 0. Then $dx = -\sin t\, dt$, and so (5.8) becomes

$$\int_{-\pi}^{0} (\cos^2 t \sin^2 t + \cos t \sin t)(-\sin t\, dt),$$

which again produces the value 4/15 on integration. $\square$

Problem 5.4   Evaluate the integral

$$I = \int_C [(2xy - z)\, dx + yz\, dy + x\, dz], \tag{5.9}$$

where (i) C is the curve $x = t$, $y = 2t$, $z = t^2 - 1$ with t increasing from 0 to 1, (ii) C consists of two straight line segments, from the origin to the point $(1, 0, -1)$ and from $(1, 0, -1)$ to the point $(2, 3, -3)$.

Solution. The notation in (5.9) denotes the sum of three separate line integrals, with respect to x, y and z respectively.

(i) Substitute for x, y and z in terms of t in the integrand, using the defining equations of C. Since we have

$$dx = dt, \quad dy = 2\, dt, \quad dz = 2t\, dt,$$

69

(5.9) becomes

$$I = \int_0^1 \{[4t^2 - (t^2 - 1)] \, dt + 2t(t^2 - 1)(2 \, dt) + t(2t \, dt)\}$$

$$= \int_0^1 (4t^3 + 5t^2 - 4t + 1) \, dt$$

$$= \left| t^4 + \tfrac{5}{3}t^3 - 2t^2 + t \right|_0^1 = \tfrac{5}{3}.$$

(ii) Let C_1, C_2 denote the line segments from O to $(1, 0, -1)$, and from $(1, 0, -1)$ to $(2, 3, -3)$, respectively. Along C_1 we have $y = 0$ and $z = -x$, and so $dy = 0$, $dz = -dx$. Therefore, the contribution to I along C_1 is

$$I_{C_1} = \int_{C_1} (-z \, dx + x \, dz) = \int_0^1 (x \, dx - x \, dx) = 0.$$

For C_2, we note that the standard form for the cartesian equations of the straight line joining the points (x_1, y_1, z_1) and (x_2, y_2, z_2) is

$$\frac{x - x_1}{x_2 - x_1} = \frac{y - y_1}{y_2 - y_1} = \frac{z - z_1}{z_2 - z_1}.$$

The line on which C_2 lies therefore has equations

$$\frac{x - 1}{2 - 1} = \frac{y - 0}{3 - 0} = \frac{z - (-1)}{-3 - (-1)},$$

giving
$$y = 3(x - 1), \qquad z = 1 - 2x, \tag{5.10}$$

and on taking differentials,
$$dy = 3 \, dx, \qquad dz = -2 \, dx. \tag{5.11}$$

Substituting from (5.10), (5.11) in (5.9), with C_2 in place of C, we get for the contribution to I from C_2,

$$I_{C_2} = \int_1^2 (-12x^2 + 21x - 10) \, dx,$$

after a little reduction. This gives

$$I_{C_2} = \left| -4x^3 + \tfrac{21}{2}x^2 - 10x \right|_1^2 = -\tfrac{13}{2}.$$

Hence
$$I = I_{C_1} + I_{C_2} = 0 - \tfrac{13}{2} = -\tfrac{13}{2}. \qquad \square$$

Line integrals of the form

$$\int_C [P(x, y, z) \, dx + Q(x, y, z) \, dy + R(x, y, z) \, dz] \tag{5.12}$$

are important in physical applications, and are frequently expressed in vector notation. Let $\mathbf{F}$ denote the vector whose rectangular component form is

70

$$\mathbf{F} = P\mathbf{i} + Q\mathbf{j} + R\mathbf{k},$$

where $\mathbf{i, j, k}$ denote the unit vectors along Ox, Oy and Oz, respectively. Then a particular value of $\mathbf{F}$ is associated with each point (x, y, z). (We call $\mathbf{F}$ a vector *point-function* or *field*.) In particular, $\mathbf{F}$ has a value at each point of C.

On C, the values of x, y and z may be expressed as functions of arc length s along the curve, measured from the initial point. The vector

$$\mathbf{t} = \frac{dx}{ds}\mathbf{i} + \frac{dy}{ds}\mathbf{j} + \frac{dz}{ds}\mathbf{k} \tag{5.13}$$

is in the direction of the tangent vector, pointing in the sense in which s increases. Further, since $ds^2 = dx^2 + dy^2 + dz^2$, the magnitude of $\mathbf{t}$ is 1, i.e. (5.13) is the *unit tangent vector* to C. If we introduce the differential of the position vector,

$$d\mathbf{r} = dx\,\mathbf{i} + dy\,\mathbf{j} + dz\,\mathbf{k},$$

it follows that we can write (5.12) in either of the scalar product forms

$$\int_{\overset{}{C}} \mathbf{F} . d\mathbf{r} \quad \text{or} \quad \int_{\overset{}{C}} (\mathbf{F} . \mathbf{t})\, ds. \tag{5.14}$$

For example, if $\mathbf{F}$ is the force on a particle P when the particle is situated at the point (x, y, z), then the *work done* by $\mathbf{F}$ when P moves along the curve C is the integral of the tangential resolute of $\mathbf{F}$ with respect to arc distance along C, and is given by (5.14).

Problem 5.5 If $\mathbf{F} = (y - 2z)\mathbf{i} + xy\mathbf{j} + (2xz + y)\mathbf{k}$, evaluate $\int \mathbf{F}.d\mathbf{r}$ along the curve C in **Problem 5.4(i)**.

Solution. We have to evaluate

$$\int_{\overset{}{C}} [(y - 2z)\, dx + xy\, dy + (2xz + y)\, dz],$$

where C is defined by $x = t$, $y = 2t$, $z = t^2 - 1$, with t increasing from 0 to 1. The latter equations give

$$dx = dt, \quad dy = 2\,dt, \quad dz = 2t\,dt,$$

and so the required integral becomes, on substitution,

$$\int_{\overset{}{C}} \mathbf{F}.d\mathbf{r} = \int_0^1 2(2t^4 + t^2 + t + 1)\, dt = \tfrac{67}{15}. \qquad \square$$

5.2 Green's Theorem in the Plane

A form of this theorem states that if R is a region of the xy plane bounded by a simple closed curve C, and if $P(x, y), Q(x, y)$ are continuous and have continuous first partial derivatives in R and on C, then

$$\int_C (P\,dx + Q\,dy) = \iint_R \left(\frac{\partial Q}{\partial x} - \frac{\partial P}{\partial y}\right) dx\,dy, \qquad (5.15)$$

where C is described in the positive sense (i.e. the anticlockwise sense when the plane is viewed from the side of positive z, so that R is kept to the left during the circuit).

When R is such that every line which is parallel to a coordinate axis meets C in at most two points, the proof is elementary. For we can regard C as consisting of a 'lower' curve $y = y_1(x)$ and an 'upper' curve $y = y_2(x)$, where $y_2 > y_1$, and x ranges over an interval $a \leqslant x \leqslant b$. Since the sense of description of C is positive, x will increase from a to b along the lower curve, and decrease from b to a along the upper curve. Thus,

$$\int_C P\,dx = \int_a^b P[x, y_1(x)]\,dx + \int_b^a P[x, y_2(x)]\,dx$$

$$= -\int_a^b [P(x, y_2) - P(x, y_1)]\,dx$$

$$= -\int_a^b \left(\int_{y_1}^{y_2} \frac{\partial P}{\partial y}\,dy\right) dx = -\iint_R \frac{\partial P}{\partial y}\,dx\,dy. \qquad (5.16)$$

By interchanging the roles of x and y, we may similarly regard C as consisting of a 'left' curve $x = x_1(y)$ and a 'right' curve $x = x_2(y)$, with $x_1 < x_2$, and prove

$$\int_C Q\,dy = \iint_R \frac{\partial Q}{\partial x}\,dx\,dy. \qquad (5.17)$$

Adding (5.16) and (5.17) gives the required result.

The theorem is valid even when R does not satisfy the condition assumed in this proof. The extension to more general *simply-connected* regions of the plane requires a closer analysis involving subdivision. (A simply-connected region R, of a surface or in three-dimensional space, is one in which every closed curve can shrink continuously to a point without leaving R. For example, in the plane, a disc is simply-connected but an annulus is not; in three-dimensional space a solid sphere is, but a torus (doughnut-ring) is not.)

Problem 5.6 Extend Green's theorem to the case where R is the annulus $b^2 \leqslant x^2 + y^2 \leqslant a^2$, assuming the theorem to be valid for any simply-connected region of the plane.

Solution. Let C_1 and C_2 denote the smaller and larger circles, respectively, which together bound R (Fig. 5.1). Let the positive sense of des-

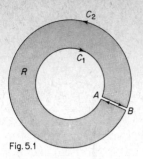

Fig. 5.1

cription of the parts C_1 and C_2 of the boundary be interpreted in each case as that in which R is kept to the left.

Introduce a 'cross-cut' AB from C_1 to C_2, to form a closed curve C_1, AB, C_2, BA which encloses a simply-connected region. Applying (5.15) to this region, we have that since the contributions to the line integral from AB and BA cancel,

$$\int_C (P\,dx + Q\,dy) = \iint_R \left(\frac{\partial Q}{\partial x} - \frac{\partial P}{\partial y}\right) dx\,dy,$$

where C denotes the sum of the curves C_1 and C_2 each described in the positive sense.

This procedure can, of course be applied to more general cases. ☐

Problem 5.7 Show that the line integral

$$\int_C [x(x^2 + y^2)\,dx + (2x + x^2 y + y^3)\,dy],$$

where C is a simple closed curve in the xy plane, is proportional to the area of the region R enclosed by C.

Solution. Let $P = x(x^2 + y^2)$, $Q = 2x + x^2 y + y^3$. Then

$$\frac{\partial Q}{\partial x} - \frac{\partial P}{\partial y} = (2 + 2xy) - 2xy = 2.$$

Therefore, by Green's theorem, the given integral is

$$\int_C (P\,dx + Q\,dy) = \iint_R 2\,dx\,dy = 2A,$$

where A is the area of R. Hence the result. ☐

Problem 5.8 If $P(x, y)$ and $Q(x, y)$ satisfy the conditions in the statement of Green's theorem (5.15), show that a sufficient condition for the line integral

73

F

$$\int_A^B (P\,dx + Q\,dy) \qquad (5.18)$$

to be independent of the curve joining A to B (provided the curve and the given points A and B lie in R) is that

$$\partial Q/\partial x = \partial P/\partial y, \qquad (5.19)$$

everywhere in R.

Solution. Let C_1 and C_2 be any two non-intersecting curves from A to B, each entirely in R. The closed curve from A to B and back to A, consisting of C_1 and $-C_2$ (i.e. C_2 reversed) encloses a region in which it is given that (5.19) holds. Hence, by Green's theorem

$$\int_{C_1} (P\,dx + Q\,dy) - \int_{C_2} (P\,dx + Q\,dy) = 0.$$

That is, the line integral (5.18) when taken along C_1 has the same value as when taken along C_2. This proves the result for non-intersecting curves. To treat the case of intersecting curves, dissect them at points of intersection and apply the above argument to the various sections.

It may be shown that the stated condition (5.19) is also *necessary* for (5.18) to be independent of the path joining A and B. $\qquad\blacksquare$

5.3 Surface Integrals A surface in space is a two-parameter locus of points defined by an equation of the form

$$z = f(x, y); \qquad (5.20)$$

or
$$F(x, y, z) = 0; \qquad (5.21)$$

or a set of equations

$$x = x(u, v), \quad y = y(u, v), \quad z = z(u, v). \qquad (5.22)$$

We shall suppose that the functions appearing here are continuous and have continuous partial derivatives in the regions in which they are defined.

A *closed* surface is one which is of bounded extent but has no boundary curve; examples being the sphere, ellipsoid and torus. Some surfaces, such as the Möbius strip (e.g. a strip of paper formed into a band with a single twist) have only one side, and are called *non-orientable*. More familiar surfaces like the sphere and torus are two-sided or *orientable*. We shall be concerned only with these.

Corresponding to (5.22), let

$$A = \frac{\partial(y, z)}{\partial(u, v)}, \quad B = \frac{\partial(z, x)}{\partial(u, v)}, \quad C = \frac{\partial(x, y)}{\partial(u, v)}. \qquad (5.23)$$

If $C \neq 0$, we can solve the first pair of (5.22), for u and v, and substitute in

74

the third to arrive at the form (5.20). Similar results apply if $A \neq 0$, or $B \neq 0$. It will be convenient to suppose that the surfaces considered can be divided into a finite number of parts, in each of which equations like (5.22) hold with A, B and C all non-zero.

The *surface integral* of $\phi(x, y, z)$ over a surface S is defined as follows. Let S be divided arbitrarily into n elements with respective areas ΔS_i $(i = 1, 2, \ldots, n)$. Let ϕ_i denote the value of ϕ at any point on the ith element. If

$$\sum_{i=1}^{n} \phi_i \, \Delta S_i$$

tends to a definite limit, as all the dimensions of the ΔS_i tend to zero and n tends to infinity, then this limiting sum is the surface integral of ϕ over S, and is written

$$\iint_S \phi(x, y, z) \, dS. \tag{5.24}$$

This always exists when ϕ is continuous.

For example, when $\phi \equiv 1$, (5.24) represents the area of S. Again, when ϕ is the area mass density of a material film forming the surface S, then (5.24) is the total mass of the film.

Problem 5.9 Show that the area of the surface S whose equation is $z = f(x, y)$ may be written

$$\iint_S dS = \iint_R \sqrt{(1 + f_x^2 + f_y^2)} \, dx \, dy \tag{5.25}$$

where R is the projection of S on the xy plane (i.e. the region in which f is defined).

Solution. Let S be divided into elements whose projections on the xy plane are rectangles of side Δx parallel to Ox and Δy parallel to Oy. If ΔS_i is the area of a typical element, located where the upward-drawn normal to S has direction-cosines (l, m, n), then by projection

$$n \, \Delta S_i = \Delta x \, \Delta y, \tag{5.26}$$

since n is the cosine of the angle between the normal to S and Oz. If we put the equation of S in the form

$$F(x, y, z) \equiv z - f(x, y) = 0,$$

then since the direction-cosines in question are proportional to (F_x, F_y, F_z), we have that they are proportional to $(-f_x, -f_y, 1)$. (See, for example, L. Marder: *Vector Algebra* (Chapter 5) in this series.) But the direction-cosines satisfy $l^2 + m^2 + n^2 = 1$, and so we must have

$$(l, m, n) = (-f_x, -f_y, 1) / \sqrt{(1 + f_x^2 + f_y^2)}.$$

Substituting for n from this equation into (5.26) gives
$$\Delta S_i = \sqrt{(1+f_x^2+f_y^2)}\,\Delta x\,\Delta y.$$
Summing over all i and taking the limit as $\Delta S_i \to 0$ gives the required result. $\qquad\qquad\qquad\qquad\qquad\qquad\qquad\qquad\qquad\qquad\qquad\quad\square$

Problem 5.10 Find the area of the surface S which consists of that part of the surface $2x^2+3y^2+z = 1$ contained within the elliptic cylinder $4x^2+9y^2 = 1$.

Solution. We note that the elliptic cylinder in question is formed by the family of lines parallel to Oz which pass through the ellipse in the xy plane: $4x^2+9y^2 = 1$, $z = 0$. Hence we require the area of the surface
$$z = f(x, y) = 1-2x^2-3y^2$$
for which (x, y) lies in the region R of the xy plane defined by $4x^2+9y^2 \leqslant 1$. By (5.25), the required area is
$$\iint_R \sqrt{(1+f_x^2+f_y^2)}\,dx\,dy = \iint_R \sqrt{(1+16x^2+36y^2)}\,dx\,dy.$$

To evaluate the integral, transform to modified polar coordinates by putting
$$4x = r\cos\theta, \qquad 6y = r\sin\theta. \qquad\qquad\qquad (5.27)$$
The image of R in the $r\theta$ plane is the region $0 \leqslant r \leqslant 2, 0 \leqslant \theta < 2\pi$, and the Jacobian of the transformation (5.27) is easily found to be
$$\partial(x, y)/\partial(r, \theta) = r/24.$$
Therefore, the required area is
$$\frac{1}{24}\int_0^{2\pi}\int_0^2 \sqrt{(1+r^2)}\,r\,dr\,d\theta = \frac{1}{24}\int_0^{2\pi}\left.\frac{1}{3}(1+r^2)^{\frac{3}{2}}\right|_0^2 d\theta = \frac{\pi}{36}(5\sqrt{5}-1). \quad\square$$

Problem 5.11 Evaluate the surface integral of $\phi(x, y, z) = 6x^2+3y^2+z$ over the surface S in Problem 5.10.

Solution. We have to evaluate
$$\iint_S \phi(x, y, z)\,dS = \iint_R \phi[x, y, f(x, y)]\sqrt{(1+16x^2+36y^2)}\,dx\,dy, \quad (5.28)$$
(using the results and notation of Problem 5.10). Now,
$$\phi(x, y, f) = 6x^2+3y^2+(1-2x^2-3y^2)$$
$$= 4x^2+1 = \tfrac{1}{4}r^2\cos^2\theta+1,$$
in terms of the modified polar coordinates (5.27). Therefore, (5.28) becomes
$$\frac{1}{24}\int_0^{2\pi}\int_0^2 (\tfrac{1}{4}r^2\cos^2\theta+1)\sqrt{(1+r^2)}\,r\,dr\,d\theta.$$

76

To simplify the r integration, make the substitution $r = \sqrt{(t^2-1)}$. Then the last integral becomes

$$\frac{1}{24} \int_0^{2\pi} \int_1^{\sqrt{5}} [\tfrac{1}{4}(t^2-1)\cos^2\theta + 1] t^2 \, dt d\theta$$

$$= \frac{\pi}{24} \int_1^{\sqrt{5}} [\tfrac{1}{4}(t^2-1)+2] t^2 \, dt$$

$$= \frac{\pi}{24} \left| \frac{1}{20} t^5 + \frac{7}{12} t^3 \right|_1^{\sqrt{5}} = \frac{\pi}{720}(125\sqrt{5} - 19). \qquad \square$$

When the equation of a surface S is given in the form $F(x, y, z) = 0$, the direction-cosines (l, m, n) of the upward-drawn normal are proportional to (F_x, F_y, F_z), and since $n > 0$, we have

$$n = |F_z| / \sqrt{(F_x^2 + F_y^2 + F_z^2)},$$

so that the surface integral of $\phi(x, y, z)$ over S is calculated from

$$\iint_S \phi \, dS = \iint_R \phi \, \frac{\sqrt{(F_x^2 + F_y^2 + F_z^2)}}{|F_z|} \, dx \, dy, \qquad (5.29)$$

though we still need to solve $F = 0$ for z as the integrand must be expressed as a function of x and y.

The case where the surface is expressed in parametric form is dealt with in the next problem.

Problem 5.12 Show that the area of the surface S:

$$x = x(u, v), \quad y = y(u, v), \quad z = z(u, v) \qquad (5.30)$$

is given by

$$\iint_S dS = \iint_\Sigma \sqrt{(A^2 + B^2 + C^2)} \, du \, dv, \qquad (5.31)$$

where A, B, C are the Jacobians (5.23) and Σ is the image of S in the uv plane.

Give a corresponding expression for the surface integral of $g(x, y, z)$ over S.

Solution. As stated on p. 75, we suppose that A, B and C are non-zero, and so (5.30) are equivalent to a single equation $F(x, y, z) = 0$. The latter is identically satisfied if x, y and z are replaced by (5.30), and by partial differentiation with respect to u (with v constant),

$$F_x x_u + F_y y_u + F_z z_u = 0.$$

Differentiation with respect to v (with u constant) gives a second, similar equation, with v in place of u. On eliminating F_x, F_y and F_z from

77

these two equations we obtain
$$F_x/A = F_y/B = F_z/C.$$
By (5.29), with $\phi \equiv 1$, we thus get for the area of S
$$\iint_S dS = \iint_R \frac{\sqrt{(A^2+B^2+C^2)}}{|C|} \, dx \, dy, \qquad (5.32)$$
(on substituting $F_x/F_z = A/C$, etc.).

Transforming the last integral by means of the first pair of equations (5.30), we replace $dx \, dy$ by $|C| \, du \, dv$, since C is the Jacobian of the transformation. The region of integration in the uv plane is the image of R under the transformation, and is therefore Σ, whence (5.31) follows.

Let $G(u, v)$ be the function obtained by substituting (5.30) in $g(x, y, z)$. Then (5.31) will remain valid if we introduce the factor g in the integrand on the left, and G in the integrand on the right, giving
$$\iint_S g \, dS = \iint_\Sigma G(u, v) \sqrt{(A^2+B^2+C^2)} \, du \, dv. \qquad \square \qquad (5.33)$$

Problem 5.13 Find (i) the mass, (ii) the centroid, (iii) the moment of inertia about Oz, of a thin material sheet of uniform density k which forms the surface S:
$$x = u \cos v, \quad y = \tfrac{1}{2}u^2, \quad z = u \sin v, \qquad (5.34)$$
where $0 \leqslant u \leqslant 1, 0 \leqslant v \leqslant \pi$.

Solution. Note that elimination of u and v from (5.34) gives $x^2+z^2 = 2y$, which represents a paraboloid of revolution obtained by rotating the parabola $x^2 = 2y$ about Oy. By inspection of the limits for u and v, we see that S consists of that part of this surface for which $z \geqslant 0$ and $0 \leqslant y \leqslant \tfrac{1}{2}$ (Fig. 5.2).

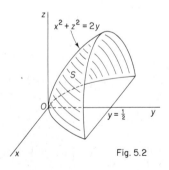

Fig. 5.2

By (5.34),

$$A = \frac{\partial(y, v)}{\partial(u, v)} = \begin{vmatrix} u & 0 \\ \sin v & u \cos v \end{vmatrix} = u^2 \cos v.$$

Likewise, we find

$$B = -u, \qquad C = u^2 \sin v,$$

whence

$$\sqrt{(A^2 + B^2 + C^2)} = u\sqrt{(u^2 + 1)}.$$

(i) The mass is

$$M = \iint_S k \, dS = k \int_0^\pi \int_0^1 u\sqrt{(u^2 + 1)} \, du \, dv$$

$$= k \int_0^\pi \left| \tfrac{1}{3}(u^2 + 1)^{\frac{3}{2}} \right|_0^1 \, dv$$

$$= \tfrac{1}{3}(2\sqrt{2} - 1)k\pi.$$

(ii) Let the centroid be the point $(\bar{x}, \bar{y}, \bar{z})$. By symmetry, $\bar{x} = 0$. Since $y = \tfrac{1}{2}u^2$ on S, we have

$$M\bar{y} = \iint_S yk \, dS = k \int_0^\pi \int_0^1 \tfrac{1}{2}u^3 \sqrt{(u^2 + 1)} \, du \, dv = (\sqrt{2} + 1)k\pi/15,$$

where we have used the substitution $u = \sqrt{(t^2 - 1)}$ to perform the first integration. Therefore, by (i),

$$\bar{y} = (5 + 3\sqrt{2})/35.$$

Similarly,

$$M\bar{z} = \iint_S zk \, dS = k \int_0^\pi \int_0^1 u^2 \sqrt{(u^2 + 1)} \sin v \, du \, dv$$

$$= k[3\sqrt{2} - \ln(1 + \sqrt{2})]/4,$$

where the substitution $u = \sinh\theta$ has been used to carry out the first integration. Then, $\bar{z}$ is determined on dividing by (i).

(iii) The moment of inertia about Oz is

$$I_z = \iint_S (x^2 + y^2)k \, dS = k \int_0^\pi \int_0^1 (\cos^2 v + \tfrac{1}{4}u^2)u^3 \sqrt{(u^2 + 1)} \, du \, dv$$

$$= k\pi \int_0^1 (\tfrac{1}{2} + \tfrac{1}{4}u^2)u^3 \sqrt{(u^2 + 1)} \, du,$$

(carrying out the v integration first). This gives

$$I_z = (5\sqrt{2} + 2)k\pi/42. \qquad \square$$

EXERCISES

1. If $\phi(x, y, z) = 2z - x^2 - y^2$, evaluate $\int \phi \, ds$ over the curve $x = t \cos t$, $y = t \sin t$, $z = t^2$, t increasing from 0 to 1.

2. Evaluate

$$\int_C \frac{x^2 y}{x^2 + y^2}\, dx,$$

where C is the circle $x^2 + y^2 = 1$ in the xy plane, described in the positive sense.

3. Evaluate

$$\int_C [xy\, dx + (z - x)\, dy + 2yz\, dz],$$

where C is formed of the three straight line segments, from the origin to the point $A(1, 0, 0)$, from A to $B(1, 2, 0)$, and from B to $(1, 2, -2)$.

4. Use Green's theorem in the plane to evaluate

$$\int_C [(x + 1)ye^x\, dx + x(e^x + 1)\, dy],$$

where C is the circle $x^2 + y^2 = a^2$, $z = 0$, described in the positive sense.

5. Find the surface area of that part of the hemisphere $x^2 + y^2 + z^2 = 16$, $z \geq 0$, whose projection on the xy plane is bounded by the curve $r = 2\theta$, $0 \leq \theta \leq \frac{1}{2}\pi$ (in plane polar coordinates) and part of the y axis.

6. Evaluate

$$\iint_S xy\, dS,$$

where S is the surface $x = u \cosh v$, $y = u \sinh v$, $z = \frac{1}{2}(1 - u^2)$, $0 \leq u \leq 1$, $0 \leq v \leq 1$.

Answers to Exercises

Chapter 1

1. $\cos y + y^2\cos xy,$ $-x\sin y + \sin xy + xy\cos xy,$ $-y^3\sin xy,$ $-\pi^3/8,$
   $-(1+\tfrac{1}{4}\pi^2).$

2. $f_{xy}(0,0) = 1,$ $f_{yx}(0,0) = -1.$

3. $k\cos k,$ $-(k+e^{-2})\cos k - k^2\sin k,$ where $k = (1+e)/e^2.$

4. $x(x^2+y^2)^{-1},$ $y(x^2+y^2)^{-1},$ $-y(x^2+y^2)^{-1},$ $x(x^2+y^2)^{-1}.$

5. $z = (1+x)(e^y-1).$

7. $xy(e^{x+y}-x).$

8. $0.42.$

Chapter 2

1. $(yz-2x)/(y+2zu),$ $(xz-u)/(y+2zu),$ $(xy-u^2)/(y+2zu).$

2. $u[(1-3v^2y)u^2+y]\,[(3u^2x-1)(3v^2y-1)-xy]^{-1},$
   $u[(3u^2x-1)-xu^2]\,[(3u^2x-1)(3v^2y-1)-xy]^{-1}.$

3. $w = u^2 - 2v.$

4. Two; e.g. $v = -tw,$ $u = t+2w.$

5. $(1+4xy)^{-1}.$

6. $[2x(z-y-x)]^{-1},$ provided $x \neq 0,\ z-y-x \neq 0.$

7. S is that part of the annulus $a \leqslant \sqrt{(u^2+v^2)} \leqslant b$ for which $u \leqslant 0, v \geqslant 0.$
   $\partial(u,v)/\partial(x,y) = e^{2x}.$ The senses are the same.

Chapter 3

1. $-10 - [8(x-1)-13(y+2)] + [(x-1)^2+13(x-1)(y+2)-6(y+2)^2] -$
   $[6(x-1)(y+2)^2-(y+2)^3] + (x-1)(y+2)^3.$

2. $1+\tfrac{1}{2}(x^2+y^2)+\tfrac{3}{8}(x^4+2x^2y^2+y^4)+\ldots.$

3. Saddle point at $(0,0)$, min. at $(-1,1)$, min. at $(1,-1).$

4. Min. at $(0,0,-1).$

5. Least is c, greatest is $[a^4+b^4+c^4]^{\frac{1}{4}}.$

6. $x = t = \tfrac{67}{69},$ $y = \tfrac{2}{23},$ $z = \tfrac{14}{69}.$

7. $4(y+z)^2+x(4y-4z-x) = 0.$

8. $x(x^2+y^2+z^2)+2y^2 = 0.$

9. $xyz = 1.$

Chapter 4

1. (i) 86, (ii) $\frac{1}{6}$.

2. $\frac{3}{35}$.

3. sinh 1.

4. $\frac{1}{60}$.

5. 24π.

7. $\frac{1}{60}$.

Chapter 5

1. $\dfrac{11\sqrt{6}}{40} - \dfrac{\sqrt{5}}{200}\ln(\sqrt{5}+\sqrt{6})$.

2. $-\pi/4$.

3. 6.

4. πa^2.

5. $8[\pi - 2\sin^{-1}(\pi/4)] - \pi\sqrt{((16-\pi^2)}$.

6. $[2\sqrt{2}(\cosh^5 1 - \cosh 2) + \sinh^2 1]/15\cosh 2$.

Index